Measuring the World, by John Austin

Published by Enigma Scientific Publications

© John Austin 2014, all rights reserved.
No part of this publication may be reproduced, stored in a retrieval system, or transmitted, in a form or by any means, electronic, mechanical, photocopying, recording or otherwise without the prior permission of the author.

First published in England in 2014 by Enigma Scientific Publications, Berkshire, UK
http://www.enigmascientific.co.uk

First print version 5 October, 2014
ISBN number 9781849145626

Createspace version 2 May 2015
ISBN number 9781502578457

Measuring the World, by John Austin

Measuring the World, by John Austin

Contents

1. Introduction — 11
   1.1 Spelling of unit names — 15
   1.2 Familiarity with units and measurements — 15
   1.3 Precision in language — 16
   1.4 Book layout — 19

2. The base and derived units — 23
   2.1 Introduction — 23
   2.2 Unit systems — 26
   2.3 Changes in primary standards — 30
   2.4 Derived units — 32
   2.5 The mass primary standard — 33
   2.6 The fundamental physical constants — 34
   2.7 Scale analysis and alternative unit systems — 37
   2.8 Practical scale analysis — 39
   2.9 Computing Units — 39

3. Time — 43
   3.1 Introduction — 43
   3.2 History of the calendar — 46
   3.3 The Gregorian calendar — 47
   3.4 The World calendar — 49
   3.5 A history of early clocks — 51
   3.6 A history of the atomic clock — 52
   3.7 Modern pendulum clocks — 54
   3.8 The second — 55
   3.9 Leap seconds — 57
   3.10 Timing in sport — 59

4. Frequency — 67
    4.1 Introduction — 67
    4.2 Units of frequency — 68
    4.3 Radioactivity — 68
    4.4 Radioactivity and human health — 70
    4.5 The dating of archaeological samples with C-14 — 71
    4.6 Dating with radioactive uranium — 75
    4.7 Potassium-argon dating — 77
    4.8 Summary of nuclides used for dating and testing — 78
    4.9 The age of the Earth — 79
    4.10 Other timing methods — 80

5. Length — 85
    5.1 Introduction — 85
    5.2 The history of units: the human body — 86
    5.3 The history of modern units — 89
    5.4 Conversion of units — 93
    5.5 Peculiar units — 93
    5.6 The electromagnetic spectrum — 94
    5.7 Navigation — 98
    5.8 Map scaling — 101
    5.9 Short and long-sightedness — 102
    5.10 Velocity — 103
    5.11 Atmospheric wind — 105
    5.12 Distance measurement in sport — 108

6. Area and volume — 115
    6.1 Introduction — 115
    6.2 Units of area and volume — 116
    6.3 The areas of small items — 117
    6.4 The area of medium-sized items: paper magic — 117
    6.5 Measuring large items – land area — 119
    6.6 Approximate conversions – area — 121
    6.7 Volume measurements of fluids — 122
    6.8 Dry volume measurement — 124

6.9 Peculiar units	126
6.10 Approximate conversions – volume	127
6.11 The Earth's changing icecaps	128

7. Mass — 131
7.1 Introduction	131
7.2 The history of units	133
7.3 The difference between weight and mass	134
7.4 Comparison between units	135
7.5 Peculiar units	137
7.6 Weighing machines	138
7.7 When your weight matters and when it doesn't	140
7.8 Allometric Scaling in Biology	141
7.9 Density and specific gravity	143

8. Temperature — 149
8.1 Introduction	149
8.2 Celsius and Fahrenheit	150
8.3 The properties of gases	153
8.4 Temperature conversions	154
8.5 The thermodynamic temperature scale	155
8.6 The design of thermometers	158
8.7 Atmospheric temperature	160
8.8 Atmospheric measurements	162
8.9 The greenhouse effect and climate change	163

9. Force — 171
9.1 Introduction	171
9.2 Units of force	174
9.3 Conversion of units	175
9.4 Approximate forces	175
9.5 Understanding gravity	176
9.6 Measuring Earth's gravity and Earth's rotation	178
9.7 The mass of the Earth	180
9.8 Gravitational forces on satellites and spacecraft	181
9.9 Tidal forces	184

9.10 Frictional forces	185
9.11 Rotational forces	187
9.12 Measuring spin rates	191
9.13 Other forces	191

10. Pressure — 199

10.1 Introduction	199
10.2 Units of pressure	201
10.3 Atmospheric pressure	201
10.4 Measuring atmospheric pressure	205
10.5 The mass of the atmosphere	207
10.6 Earth's rotation: winds and bathtubs	209
10.7 Water pressure and subterranean pressure	212
10.8 Sound waves	215
10.9 The speed of sound	217
10.10 The Doppler effect	218
10.11 Sound waves in solids: earthquakes	219
10.12 Tyre pressures	222
10.13 Engine exhaust pressure	223
10.14 Pressure calculations with fps units	224

11. Work, energy and power — 231

11.1 Introduction	231
11.2 Mechanical versus heat energy	233
11.3 Conversion between units	234
11.4 Dietary energy	235
11.5 Power units	236
11.6 Energy expenditure of exercise	237
11.7 Very large energy units: bombs	240
11.8 Natural phenomena	241
11.9 The performances of aircraft and rockets	243
11.10 Energy content of fuels	245
11.11 Simple machines	246

12. Electrical current and electrical power   253
    12.1 Introduction   253
    12.2 Electrical current, voltage and resistance   254
    12.3 Capacitance   256
    12.4 Inductance   256
    12.5 Measuring electrical properties   258
    12.6 Magnetism   259
    12.7 Societal uses of magnetic fields   260
    12.8 Natural electricity   263
    12.9 Power stations   264
    12.10 Nuclear power: fission   266
    12.11 Nuclear power: fusion   267
    12.12 Renewable power   269
    12.13 Off-peak power use   271
    12.14 Energy efficiency in the home   271

13. Amount of substance   279
    13.1 Introduction   279
    13.2 The periodic table and atomic weights   281
    13.3 The ideal gas law   285
    13.4 The number of atoms per mole   288
    13.5 Other molar measures: acidity and catalysis   289

14. Light brightness   293
    14.1 Introduction   293
    14.2 Reflection of a light beam   295
    14.3 Refraction of a light beam   295
    14.4 Diffraction of a light beam   296
    14.5 Moths to a flame   298
    14.6 Luminous intensity and brightness   299
    14.7 Measurements with light meters   300
    14.8 Light bulbs   301
    14.9 Light emitting diodes   302
    14.10 Calculating comfortable light levels   304

15. The very small and the very large — 309
    15.1 Introduction — 309
    15.2 The very very small — 310
    15.3 Mass and energy of sub-atomic particles — 311
    15.4 Microscopes for the very small — 315
    15.5 Microscopes for the very, very small — 316
    15.6 Cosmic rays — 316
    15.7 The very large — 319
    15.8 Observing the universe — 321
    15.9 The Doppler effect and the age of the universe — 323
    15.10 The sun and other stars — 324

16. Unit agreement — 335
    16.1 Introduction — 335
    16.2 Changing from fps to metric units — 337
    16.3 The problems of mixed units — 340
    16.4 The cost of becoming completely metric — 341
    16.5 Major problems due in part to mixed units — 344
    16.6 National efforts towards metrication — 348

17. The future of the metric system — 355
    17.1 Introduction — 355
    17.2 The kilogramme — 356
    17.3 The fundamental constants of the SI system — 359
    17.4 Adoption of the new units — 364

18. Future unit use — 367

Acknowledgements — 374

Appendix A Definition of terms and derived units for the metric system — 375
A1 Metric multipliers — 375
A2 Definition of the SI base units — 375
A3 List of SI units — 376
A4 Definition of common terms with SI unit — 378

Appendix B Conversion tables — **381**
B1 Length — 381
B2 Area — 381
B3 Volume — 382
B4 Mass — 383
B5 Temperature — 383
B6 Other quantities — 384
Appendix C Special SI units in honour of famous scientists — **385**
Appendix D Ten units which should be confined to the dustbin of history — **389**
Index — **391**
About the Author — **401**

Measuring the World, by John Austin

Measuring the World, by John Austin

1.Introduction

Many ancient civilisations, including the Romans and Egyptians, have found the need for making measurements of various kinds. In each case, this led to the establishment of unit systems for mass, length and time. Millennia later, we have the same need for units for trade and technology, but our requirements have become somewhat more exacting. Moreover, with these activities now taking place globally, a single unit system is desirable. Accordingly, about 90% of the world population has adopted the metric system, but about 10% stubbornly refuse to be part of it and continue to use units such as the Imperial System of Units. This book is a celebration of the metric system, arguably one of mankind's major intellectual and cultural achievements. It is also intended as an encouragement to the 10% at least to try the units for size. You may not like them at first, but in time I would hope that you would see their beauty. If nothing else, see how much easier routine calculations become in the metric system, by reading this book.

Britain is a mixed-up place for units. I'm often a walking conversion table for my wife from Brazil. We have miles and miles per hour, and pounds and stones, but litres and kilogrammes and metres. The USA doesn't want anything to do with the metric system, presumably for ideological reasons. Anything thought up by foreigners can't be any good! The USA apparently prefers to lose the odd satellite because of incorrect

Measuring the World, by John Austin

units, rather than use a global system of units based on powers of ten. Their neighbours to the north have converted mostly to metric, although the process has not been completed, leaving a bit of a mixture of units in places. One of the arguments I frequently encounter is that people don't like to use the metric system because, for example, a hot day (100 °F) doesn't *sound* as hot in Celsius (37.8 °C). I tend to agree. But to most continental Europeans, 40 °C (104 °F) sounds and feels very hot! Even 35 °C is hot in Britain, as it has never exceeded 40 °C. Often, then, people convert the units but not the concept, which would be considered poor in translating a message in a foreign language. It is the idea that needs translating, not the words. There are countless examples similar to the `hot day' concept. Many people of course have been entirely educated in the metric system, and my wife from Brazil is just one of them. I don't disguise my frustration with the 10% of the world that doesn't automatically use metric. Many of them may struggle to remember how many centimetres are in a metre, or metres in a kilometre, or indeed how large a metre is in practice. I don't know if the cost has been quantified, but I'm sure that misunderstandings about units must be some sort of impediment to business. Of course as in all cultural exchanges, use of different units can bring out a subtly different cultural aspect. For example, for most people, a sunny day at 38 °C would be considered very warm and it would perhaps only be considered exceptional if the temperature reached 40 ºC. So changing units subtly changes our concepts. Overall, though it is a small price to pay. On a slightly different note, miscommunications often arise on the conversion of metric to imperial, or the converse. For example, I learnt the other day that kangaroos were often as tall as 8 ft 2 in. I might think on that basis, that a typical kangaroo might be in the range, say 7 ft 8 in to 8 ft 2 in. However, 8 ft 2 in is the conversion from the approximate 2.5 m. So, working in metric, one might guess that kangaroos are in the range 2-2.5 m. In other

Measuring the World, by John Austin

words, in one set of units the range is expected to be about 50 cm, in the other set of units the range is more like 6 in or 15 cm, quite a different value. Of course the uncertainty is a guess on my part. In all measurements we ideally want to specify the estimate of the range, or the uncertainty in a given measurement. If it is not given explicitly, as in the kangaroo example, it is often understood subconsciously and this is often conveyed in the apparent precision of the measurement. So the process of conversion of units can easily convey the wrong message if care is not taken.

Naturally, there is a one-off cost associated with converting completely from one unit system to the other, but once that expense is made, the benefits can then accrue. We will see in this book numerous examples where the benefits can continually arise, although it is often a subjective matter to quantify such benefits. For example, the metric system will open up new possibilities for mental arithmetic calculations that would be unwieldy in the Imperial System of Units. To whet your appetite for what comes, I ask the simple (but perhaps useless!) question (Chapter 6) of how many sheets of typing paper (A4) would you need to cover an acre of land? Most people would need a pocket calculator to find out, and it would likely take a few minutes. The equivalent question in the metric system (sheets of paper per hectare) can be answered in seconds with mental arithmetic alone. Other examples are strewn around the text. If you have trouble appreciating the overall benefits of the metric system, just think of me as a lazy scientist. I can't be bothered to incorporate every conversion factor. Instead I prefer to multiply factors of 10!

Let me say without hesitation that I believe the metric system (based on the metre, kilogramme and second – mks) is the best system of units ever devised by humans. I would like to see it thoroughly adopted by all humanity, including the residents of UK and USA. There are some truly beautiful connections between different aspects of the system that invoke the same emotions as a fine work of art! Take the paper sizing of Chapter 6, or physicists' drive to reduce all units to 'universal' standards (Chapter 2). In other words, if we were to contact an alien life form, we could let them know how large our units are to a very high precision. By comparison, we still wouldn't be able communicate with said aliens the difference between left and right [1]. I know it sounds esoteric, but practical devices such as the Global Positioning System (GPS) tracking, have emerged from the drive for higher reproducibility of units. Many people are probably quite unaware of the subtleties of our measuring systems, but they deserve to be more widely known. I hope that this book can go some way towards redressing the balance.

In contrast to the metric system, the Imperial system was derived from the middle ages onwards, when the underlying physics was poorly understood. There are differences in the two sets of Imperial units that have become established on either side of the Atlantic, particularly for volume measurement. Because of confusions over volumes, US volume units are not as clearly related to mass units as in the metric system, where 1 m^3 of water has mass 1000 kg to high precision. Often, as new science emerges, the Imperial system doesn't have a unit available, and so it has borrowed completely from the metric. Electrical units fit this category: all unit systems use the amp, volt etc. Before energy was properly understood, both metric and Imperial systems had separate units for heat energy (calorie, therm). The latter is still part of the Imperial system, but although the former is still used for dietary purposes, in the metric system, heat is now recognised as just another form of energy, and is given in joules.

1.1 Spelling of Unit Names

Strictly, a unit comes with a name and that name should not be changed, regardless of the cultural proclivities of individual nations. See for example [2]. A *litre* is a *litre* which is not a *liter*. My

Measuring the World, by John Austin

editorial tendencies make me uncomfortable with the latter. The SI (Systéme Internationale) that we now use, was established by the French and we should not misspell the units they established any more than we should misspell the name of its capital city Paris. Even scientific journals in the USA misspell unit names (meter instead of metre; liter instead of litre, gram instead of gramme). Journalists often prefer 'metric ton' to tonne, thus using two words when one will suffice. The USA seems to be ideologically opposed to words ending in a double letter and e --- it just looks too French for their taste, I suppose! I think that accuracy in the spelling of unit names is similar in principle to the need for precision in the use of 'measurement language', as expounded in §1.3. It seems that both kilogramme and kilogram are acceptable spellings internationally, and acceptable in the UK, but not by me.

1.2 Familiarity with Units and Measurements

There is a further aspect of units regarding their familiarity. Most of the time, we need only a general indication of the sizes of objects, although it is reassuring to know that business is required to deliver exactly what it says it will. In other words, it's important that when you buy a container of milk marked as 2 L, it does indeed contain that much. Nonetheless, most people have only an approximate idea of units and their sizes. For example, if I were to mark out a distance of 50 m on a field (but not a sports field which would have visual clues) and asked people what they thought the distance was, I would likely get answers in the range 25-100 m (or the equivalent in the Imperial system). If they can potentially make a random error of about a factor of 2, why can't they accept a smaller but systematic error of about 10%? In other words, if you don't like metres, think of them as yards instead and you wouldn't be far wrong. Similarly if you were given 500 g of apples, would you really know the difference from a pound (454 g) without weighing them first? Interestingly, the continental Europeans have retained the *livre (Fr.* translation 'pound') to mean 500 g, and in many cases it may be more useful for grocery shopping than the kilogramme.

On a related topic, beware of closet Imperial System users! It is common for science fiction films (made in the USA) to sound more exotic with the use of 'foreign sounding' units, namely metric! To do this, they simply convert units. For example, a comfortable room becomes 21 °C, yet almost nobody sets their thermostat to 21 °C. More likely, a metric user will set the thermostat to 20 °C, a round number. The 21 °C of course arise from a conversion from 70 °F, which is the sort of number someone from the USA would use. 'Star Trek' incidentally shows that there is hope for mankind as the Imperial System of units has been abandoned by about the 24th century, if not before. I was amused, though, that masses in at least one episode were measured in 'metric tons', not exactly the language of a true metric unit supporter. There will, it seems, still be some closet Imperial System users!

1.3 Precision in Language

An important issue related to units is the need for precision in language when describing quantifiable processes. For example, faith healers if you listen to them, tend to use words like 'force' or 'energy' without defining them. It can sound meaningful, impressive or knowledgeable but on examination can be vacuous and meaningless. These terms and others used in the process of measurement, are here defined as needed and summarised in Appendix A. None of the concepts are difficult, and are readily taught in high school, so how difficult can that be! Other words from scientific reports are picked up by the media and other 'influential' people. The words are then abused and the

Measuring the World, by John Austin

misunderstanding is propagated. Let's explore several of these howlers.

Quantum has been precisely recognised in physics, for over 100 years since quantum theory (the theory of the very small) became established. A quantum step in physics represents the *smallest* change, for example in the energy level of an atom. However, in recent decades, or perhaps just in the last 10 years, the media has misunderstood the terminology and now a 'quantum leap' in understanding some problem is a major step. This is virtually the opposite of that understood by scientists for the last 100 years.

A second example is that of exponential growth. An exponential rate of change has a precise meaning to a scientist. It means that the rate of change itself increases steadily with time. In fact this is a common process in nature and implies that a process, *a*, increases as given by $a = a_o e^{kt}$, where a_o is the process at time 0, *t* is time, *k* is a constant, and *e* is the base of natural logarithms (e = 2.71828....). For example, consider a simple model of reproducing bacteria in an environment with no enemies and limitless food. Suppose 2 bacteria start off and they double every 10 minutes. After 10 min., there are 4; after 20 min. there are 8; after 30 min. there are 16 etc. So the increase in number in each 10 min. period itself increases. In the first 10 min. the increase is 2, in the second 10 min., the increase is 4, and in the third 10 min., the increase is 8. In this case, the total number of bacteria at time *t* (units of 10 min.) are $n = 2^{t+1}$, or on rearrangement, $n = 2e^{kt}$ where $k = \log_e 2 = 0.6931$ (approx.). So, the process is true exponential increase, and viewed on an hourly time scale, this process is very rapid. This is consistent with the colloquial meaning of 'exponential'. Now consider another process, the cash in a savings account, subject to a 1% annual growth (a reasonable amount at the current time). The total sum, *S* in an account after *t* years would be $S = S_o 1.01^t$ where S_o is the initial sum. Again this is exponential growth $S =$

Measuring the World, by John Austin

$S_0 e^{kt}$ where $k \approx 0.01$, giving a doubling time of about 70 years. So the process is hardly rapid! How fast a process is depends on the value of k. To a scientist, both processes are exponential, but to a journalist, the latter is probably not. I have seen an increasing abuse (perhaps exponential growth?!) of the term 'exponential' in recent years, but I was somewhat surprised to see 'exponential' misused in a book published as long ago as 1989 [3], and which illustrates graphically the dangers associated with the use of mixed units (Chapter 16), i.e. combining the use of units from the two separate systems (metric and imperial). In describing the take-off of a fully-loaded jet aircraft, the subject of the book, the authors describe the take-off speed as increasing exponentially. In fact with constant thrust from the engines, the aircraft speed on the runway would have increased proportionally with time, in accordance with Newton's laws of motion. I therefore levy some criticism at journalists for introducing incorrect terms for the apparent purpose of producing more creative writing. In fact, virtually every time I read accounts of events written by non-scientists, I need first to understand what they are trying to say, and then to convert mentally to the proper language. Communication should certainly not be like that. Words should mean what they say, no more no less. I am reminded of part of a dialogue by Lewis Carroll [4]:

"When *I* use a word," Humpty Dumpty said, in rather a scornful tone, "it means just what I choose it to mean—neither more nor less."
"The question is," said Alice, "whether you *can* make words mean so many different things."

A final phrase worth questioning is "PIN number". Although not part of the measurement process, the phrase does appear in a technical environment and it hurts to feel the abuse of

English every time I hear the phrase. PINs (Personal Identity Numbers) were introduced a few decades ago when technology evolved to allow cash withdrawals using a cash card. The numbers have proliferated. Of course "PIN number" is a horrible tautology, the word "number" appearing twice. To confuse people and perhaps make my point, I sometimes call my PIN a "PI number", but the people listening are often a bit slow-witted and don't understand my point.

1.4 Book Layout

In the Chapters which follow, examples will be given from everyday life and popular science to show how practical calculations are often simpler in the metric system. Most of the Chapters contain several full colour images, which are here reproduced only in black and white. To obtain copies of the original images go to the credited source in each of the figure captions or to the page on the author's website
http://www.enigmascientific.com/ebooks/MeasuringtheWorld

In Chapter 2, broad aspects of the metric and Imperial systems are discussed. We look at physicists' attempts to achieve high precision 'universal' standards for the units, i.e., the units do not depend on artefacts such as metre rules kept under special laboratory conditions.

Chapters 3-8 cover the base units of the system, followed by their derived units. Chapter 3 covers time, Chapter 4 covers frequency, Chapter 5 length, Chapter 6 area and volume, Chapter 7 covers mass and Chapter 8 temperature. In each Chapter, a brief history is given when appropriate, and conversions from the metric system to the Imperial system are discussed. The Imperial System has numerous units, which can be most politely described as 'peculiar'. An explanation of the units is included, as appropriate. For example, what is a barrel of oil or a bushel of corn? My suspicion is that few people outside the relevant industry have anything other than the slightest inkling.

Chapter 9-13 also cover derived units. Chapter 9 covers Force, a rather important process. Chapter 10 discusses Pressure, directly derived from Force. Chapter 11 covers Work, Energy and Power. Chapter 12 on electrical energy introduces a whole new range of units. Chapter 13 is a more specialised subject discussing the amount of substance in terms of the number of atoms and molecules present. It is fundamental to our understanding of chemistry. Chapter 14 on luminosity is also related to the Energy Chapter.

The physics of the very small (e.g. the Higgs boson) and the very large (e.g. the rate of expansion of the universe) are often the subject of popular newspaper articles. Some specialist units are needed, as discussed in Chapter 15. Sometimes, some amazing howlers appear in print due to a misunderstanding of the units used. A time magazine article had the Higgs boson mass a billion times its true mass!

Chapter 16 discusses the rate of adoption of the metric system worldwide. While there have been many success stories, the sorry state of affairs in the UK and USA is revealed. 2015 is the 50th anniversary of the UK's commitment to the metric system. It's enough to make a 50+ year-old weep! The costs associated with complete adoption of the metric system are explored. Failure to convert units correctly are not always factored into the equation.

In Chapter 17, I discuss the radical proposals to redesign the metric system. Is the end of the prototype kilogramme upon us, to be replaced finally by a universal standard? In Chapter 18, I gaze into my crystal ball (refractive index 2.42 – I wish!). Will the USA be dragged kicking and screaming into the 20th (sic!) century? In many ways, the UK is not much better. Read on to find out.

Measuring the World, by John Austin

References

[1] "Galactic challenge: How would you teach left from right to an alien civilization", by David Castelvecchi, Scientific American, http://blogs.scientificamerican.com/degreesoffreedom/2011/08/28/handedness-galactic-challenge/, August 2011, Accessed 9 February 2014.
[2] Pat Naughtin, Spelling metre or meter, 2008, http://www.metricationmatters.com/docs/Spelling_metre_or_meter.pdf, accessed 15 February 2014.
[3] "Free fall", by William and Marilyn Hoffer, Grafton Books, 1989.
[4] "Through the looking glass", Lewis Carroll, 1872.

Measuring the World, by John Austin

Measuring the World, by John Austin

2. The Base and Derived Units

2.1 Introduction

All measurements are now made with metric base units, or other units such as Imperial Units which are derived directly from metric. The position was not always like that: over the centuries even after the time of Newton, units evolved in a perhaps haphazard manner as needed by industry or science of the time. The science of metrology developed as the precision of measurements and the understanding of science improved. By the 19th century, two sets of competing unit systems were still in place. The first was the Imperial System of units[1] also known as Avoirdupois, some of the units of which dated to the 14th century. The second system was the metric system[2] adopted by France in 1795. By the 19th century, units of both systems were defined in terms of artefacts kept in two main laboratories in Greenwich (London) and Paris (France). Paris retained the metre as the distance marked on a platinum rod, and the kilogramme as the mass of a platinum-iridium alloy cylinder. London retained the standard yard and the pound mass, both kept independently of the metric units. The rotation of the Earth was used for the time base unit, the second, which in both unit systems was defined as 1/86400 of the mean rotational period of the Earth.

The establishment of carefully produced units in the form of engineered artefacts supported the notion of unit

Measuring the World, by John Austin

'systems'. On the side of the pound was the 'fps' (short for foot, pound, second) system, while on the metric side was the 'cgs' (centimetre, gramme, second) or 'mks' (metre, kilogramme, second) systems. Artefacts have had a problem in that as measurement precision improved, so they have been found to change. In 1950, the Imperial yard, made of brass, was found to shrink relative to the metre at a rate of about 2 parts per million over 50 years[3]. The international prototype kilogramme has been found to drift downwards at about 5 parts in 10^9 per decade[4] relative to national copies. The Earth's rotation period is also not constant: recent estimates[5] indicate that the day is lengthening at the rate of 1.7 ms per century (2 parts in 10^9 per decade). That is in addition to the day to day variation of the length of the day (by a few hundred μs) due to the changes in major weather systems. The current concerns with the kilogramme are that the international prototype may have acquired dust on its surface over long periods. Overall, such problems favour the move away from artefacts as the primary standard towards the use of universal standards. The idea then is that the base units can literally be reproduced anywhere in the universe.

By the 1960s[6] the fps system had lost its independence: in other words, every fps unit became defined in terms of the equivalent metric one. For example, 1 inch = 2.54 cm (exactly), and 1 foot = 30.48 cm (exactly). Philosophically, this means that fps unit users are closet metric users! One of the main advantages of a unit system is that, if all the inputs of a calculation are from within the same unit system, then all the outputs are from the same system. For example, working in the mks system, consider the acceleration of a loaded aircraft, as it travels along the runway for take-off. We can use Newton's law $F = ma$ (Chapter 9), where F is the force, or engine thrust, m is the total mass of the aircraft, and a its acceleration. Take F = 400,000 N, m = 100,000 kg. So the acceleration is a = 400,000/100,000 = 4. The mks unit of force is the Newton (N), kg is the mks unit for mass, and m s^{-2} is the mks unit of acceleration. Hence the acceleration at take off is 4 m s^{-2}. An important point is the absence of any multiplying factors: the final 4 is the ratio of the two numbers.

The above calculation would be just as easy in the cgs system, except that the unit sizes are smaller, so the numbers become correspondingly larger. The equivalent calculation would be F *(dyne)/m (g)* = $4 \times 10^{10} / 1 \times 10^8$ = 400. But acceleration is now in cm s^{-2}, so the answers are the same. In the fps system, the engine thrust would need to be expressed in poundals, and the weight of the aircraft would need to be in pounds. There lies part of the problem, because the aircraft mass may just as well only be known in tons (2240 lb), or even short tons (2000 lb). In either case, a conversion factor of 2.24 or at best 2.0 is still needed. Contrast the metric system, where if the aircraft weight is known in tonnes, then it is trivial to convert to kg, since 1 t = 1000 kg. (1 t = 1 ton very closely). It is of course possible in metric to mix systems, taking one unit from cgs and one unit from mks. That would be a mistake leading to an error of several orders of magnitude (i.e. several factors of 10). The first step in any calculation, then, is to convert the units to the correct one for the system being used. In the above example, it was pointed out that the mass of the aircraft should be in kilogrammes, not tonnes. As in this example, the cgs system is not generally as practical as the mks system. Consequently, the committee on metric units, the Système International d'Unitées (SI)[7], fully endorse the mks system. They are often referred to as SI units, and from here onwards they will be largely referred to as such. Over the last 40 years, they have become very good friends of mine as well! The SI system is a comprehensive set of units to cover all measurable properties of physical objects and processes. There remains some confusion in the use of fps units, with two versions available --- Imperial ('British') and American versions. Many other specialist units are adopted by industry. These differences are explored in the individual chapters which follow. In particular, the costs to industry and society when, frankly, antiquated units are used.

2.2 Unit Systems

Once a unit system is in place, be it fps, cgs or mks, other units follow as explained in § 2.4. However, fps does not have independent electrical units, so it is an incomplete system. Instead, it borrows from cgs or mks, producing a totally mixed up combination.

In the metric system, main units are modified by the use of multiplying prefixes. For example, millimetres = mm = 10^{-3} m. This is one of the major strengths of the metric system, since it means that if a particular unit is too small or too large for use in a practical problem, then choosing a suitable prefix can adjust the values to within the desired range. In the fps system, units are preselected. For example, large distances can be measured in miles, and short distances in inches, but the only smaller unit available is the 'thou' (0.001")[8] which is not in common use. In the USA, 0.001 in is referred to as one mil. The fps system of course suffers in that it is not straightforward to convert from small to large units, as the factors involved are not powers of 10. Units such as square metres (area), cubic metres (volume) are considered to be best represented using power notation m^2, m^3 etc. Units such as square kilometres could be written as $(1000\ m)^2$ = $1 \times 10^6\ m^2$ = $1\ Mm^2$. However, this is not recommended, and when a term such as km^2 is used, it is understood that the power refers to the premultiplier as well as the base unit. In other words km^2 means square kilometres and μm^2 mean square micrometres. This is less of a problem in the Imperial system of units, as the words square and cubic are used in full or abbreviated: sq., c. In any case,

Measuring the World, by John Austin

the Imperial system is much less general than the metric system in which we can represent units such as m^4 or m^5 etc., if those were needed. Negative powers are more elegant and less ambiguous in complex combinations of units: $J\ kg^{-1}\ K^{-1}$ instead of J/kg/°C and $kg\ m^{-3}$ instead of kg/m^3 are preferred. Also acceptable is km/h for speeds but certainly not kph.

The SI system has 7 *base* units: the metre (m, Chapter 5), the kilogramme (kg, Chapter 7), the second (s, Chapter 3), the kelvin (K, temperature, Chapter 8), the ampere (A, Chapter 12), the mole (mol, amount of substance, Chapter 13) and the candela (cd, luminance, Chapter 14). These are supplemented with other units *derived* from the base units: for example, the newton (N), for measuring force (Chapter 9), the joule (J), for measuring energy and the watt (W), for measuring power (Chapter 11). Electricity (Chapter 12) requires more derived units, including the volt (V), for measuring electrical force, the henry (H), electrical inductance (Chapter 12) and the tesla (T), for magnetic field strength (Chapter 12). The definitions and details of all the SI units both base and derived, are included in the individual chapters, and in Appendix A. There is no logical equivalent in fps.

As noted in the Introduction, one of the major strengths of the metric system is the ability to set up more practical combinations of units for particular measurements, based on a set of prefixes (or multipliers) to the main units. The full list of multipliers is given in Appendix A. The SI has indicated a preference for powers of 10^3, given by multipliers k (10^3), M (10^6), G (10^9) etc. as well as m (10^{-3}), μ (10^{-6}), n (10^{-9}) etc. The advantage of multiples of three is to reduce the chances made in manual calculation. For example, if the calculation involves multiplying several terms, the first task is to reduce the problem to SI units before doing any computation. If there are only multiples of 3 in the powers of 10, the calculations are easier. The SI could equally have recommended powers of 2 or 4, but the former would not save much effort, and the latter would probably leave us with awkward numbers. For example, a distance 32135 m is more meaningfully written as 32.135 km than, say, 3.2135 x 10^4 m. The recommendation reflects our tendency towards base numbers such as thousand, million, billion etc. Incidentally, there used to be a time when the UK billion was 10^{12}. However, the use of billion in US financial and political circles to mean 10^9 has led to our adoption in the UK. Consequently, I use the US terms, fully expecting the UK terms to become extinct in the interests of better global communication. Cynic though I may be, I wonder how many people even in the US power industry, correctly understand a global-sized energy unit, the quad. This is 10^{15} BTU (about 10^{18} J), which is the the annual output from about thirty 1 GW power stations. For example, Drax in the UK, has a capacity of almost 4 GW, and supplies 7-8% of the UK demand[9].

In the US numbering system, 1 million is 10^6, a billion is 10^9 etc. So, an X-illion is 10^{3x+3} where X in the name is some Latinised version of 2,3,4 etc. This is a bit illogical to me and by comparison, the British version would have been 10^{6x} for X-illion, So, a quadrillion would have been 10^{24}. I bite my tongue and say no more about this. I suspect that most Americans would think a quadrillion is 10^{12}, although they usually call this a trillion. Quad is of course short for quadrillion therms, but it would be better if US industry used the EJ (Exajoule, 10^{18} J), and then ambiguities over the use of multipliers would not exist, and the unit would be more easily computable from power consumption in kWh (kilowatt hours, Chapter 11). Overall, I don't like the use of special names for high powers of 10. If somebody tells me that a number is so many octillion, even ignoring the above possible ambiguity and that the person doesn't understand the numbering system properly, I still have to do some mental arithmetic, to work out that the number in question is several times 10^{27}. To my mind, the power notation is

Measuring the World, by John Austin

a much more meaningful and useful way of conveying such numbers, and I will generally use that method in this book. Of course very large numbers like this are often from the domain of astronomy, the wonders of which are often conveyed to the public through journalists. Perhaps the numbers become distorted in the process. Very large numbers are also starting to enter the public consciousness through the computer industry. I used to work for a lab with a major computing installation, which measured its total disk capacity in PBytes (10^{15} Bytes), although actually in computing a kByte is often 1024 Bytes (although see § 2.8) since this is an easier number to work with (being 2^{10} in a binary environment), all the way through to higher powers of kBytes. So 1 PByte is 2^{50} *Bytes*. Anyway I never heard it referred to as multiples of quadrillion bytes. Hopefully, such terms will never be used, and the metric pre-multipliers will probably continue to be a source of new unit sizes as computing power advances. The public at large are of course already becoming familiar with TBytes (10^{12} Bytes).

An example of the use of multipliers is the calculation of the acceleration of the loaded aircraft mentioned earlier. The force is 400 kN from the engines. Now suppose fuel weighs 50 t and there are 250 people on board weighing 100 kg each, including their baggage. If the unladen aircraft weighs 25 t, the total mass of the laden aircraft is *aircraft + fuel + people* = 25 t + 50 t + 250x100 = 100x10³ kg (100 t). So the acceleration is 400x10³/(100x10³) = 4 ms^{-2}, as before (or about 0.4 x the acceleration of gravity). In this example, we have worked with the people mass in kilogrammes, and combined that with the aircraft and fuel weight in tonnes, remembering to convert everything to the SI unit (kg) before doing the actual arithmetic.

Care needs to be taken with the metric multiplier and unit names, which are all case sensitive. Lower case m is used for both milli (10^{-3}) and metre. Although potentially ambiguous, I have not known a real problem with this. It is much more important to avoid using M, except when signifying 10^6. It is nice also to see metric multipliers for money, e.g. £10k, which beats 'grand', which is somewhat dated now. Note also the difference between k (kilo) and K (kelvin, thermodynamics temperature). Even scientists are sometimes a bit sloppy on this one, at least in their typed articles.

2.3 Changes in Primary Standards

As physics and technology have developed, there has been an increasing need for higher precision measurements. It has at times been important to convert such measurements into absolute masses, lengths or times. This has necessitated revisions to the primary standard, i.e. the standard of measurement which defines the base unit. The metre itself has undergone many changes, as described in Chapter 5. For example, if the metre were defined as a distance marked on a metal bar, such a distance might be known only to one part in 10^7. Now suppose that we tried to measure a wavelength transition of a particular atomic state, for example, krypton, a gas capable of being produced at extremely high purity. We count, using electronic samplers the distance between two points, which for argument sake are 1 m apart exactly, and we find that there are 1,650,763.73 wavelengths with an uncertainty of 0.01. Then the distance is only known to 1 part in 10^7 (the precision of reproducing the original metre standard), but we can make the measurement of wavelengths to 6 parts in 10^9, that is, the error in the length, 0.01 divided by the number of wavelengths, $0.01/1.65 \times 10^6$. So our ability to measure distance using wavelengths is about 15 times better than our ability to measure distance using the points on a bar. Now, suppose we no longer use the bar, but redefine the metre as the above number of wavelengths. Then we would be able to reproduce the metre to the 6 parts in 10^9 precision. The result is that the metre has not changed to any measurable extent, but is now available to higher precision for more demanding industrial applications. Although the definitions of the primary standards, such as the metre have become awkward-looking, and sound obscure, by choosing physical properties from elements that we can measure exceptionally well, the primary standards can be defined to exquisite precision. Changes to the metre did not finish there: as described in Chapter 5, the development of lasers and high precision timing allows the metre to be defined even more precisely. So in fact the metre primary standard has passed through the following stages: Earth → metal bar → wavelength of light → velocity of light. In that process, exceptional care has been taken to ensure that its length has not altered: the only change has been the precision to which it can be reproduced.

Changes in the base unit have therefore arisen from improvements in measurement precision and a switch to universal standards. Although the unit system may now have converged for the metre, it depends on the universal speed limit equal to the velocity of light in a vacuum, c. Ironically, recent apparent velocities in excess of c were measured for neutrino fluxes, after the last revision of the metre standard, but the apparent super-luminary velocity turned out to be a mistake in the timings[10]. Nonetheless, possible improvements in the units may yet take place, benefiting from future high precision measurements. Apart from higher precision measurements of experiments that are currently established, improvements in the measurements of the physical constants can have a major bearing on the unit standards. As explained in § 2.5, the kilogramme is expected to be shortly revised, and this is one of the most exciting events in modern metrology, as it will likely trigger a major change in the definition of the primary standards, as further explained in Chapter 17. Since the fundamental physical constants are so important to science and industry, these constants are discussed in § 2.6.

Measuring the World, by John Austin

2.4 Derived Units

Many important units are derived from the base units, and the underlying physics. For example, using Newton's law $F = ma$, relating the force to an object of mass m undergoing an acceleration a, when a is in m s^{-2} (an acceleration in SI) and the mass is in kilogrammes, kg, then the force is in newtons, N. An example of this was given in § 2.1. Similarly, energy is the work done against a force F, determined as a product of the force and distance moved. For example, a person who is exercising by lifting a mass (m) of 50 kg from the floor to a height of 1.5 m (h) does work of amount mgh where g is the acceleration due to gravity (9.8 ms^{-2}). Work done = 50x9.8x1.5 = 735 J. The work, or energy expended is in joules (J), the SI unit for energy. This energy is produced by chemical reactions in the muscles requiring the consumption of food. Exercise and diet are discussed from the energy perspective in Chapter 11. The joule, then, is the product of a force in newtons and distance in metres, or 1 J = 1 N m. As discussed further in Chapter 11.

The above calculations can of course be done in the fps system, providing the appropriate multiplying constants are included. For example, energy is indicated strictly in foot poundals and power in foot poundals per second. The poundal is often ignored, with pound force used as a force. This is often confusing as there is sometimes a g factor (32.2 ft s^{-2}) that needs to be included. For example, 1 horsepower (hp) is often defined as 550 ft lb s^{-1}, but there is a hidden g factor, so it is more carefully written as 550 g ft lb s^{-1}. Energy and power units of the SI, especially watts for power, are often better understood even in those countries that otherwise ignore the metric system.

With the base unit of the ampere defined, electrical units volts (V) and ohms (Ω) and other electrical units follow, as explained in Chapter 12. These are also used by the fps system as there is no equivalent. This means that although the SI connects successfully between electrical and mechanical power and energy, the fps system is forced to adopt conversion constants. See Appendix A for the definition of terms and the name of derived units in SI.

2.5 The mass primary standard

The mass of the prototype kilogramme can 'only' be compared to a precision of about 2 parts in 10^8[11] whereas distances can be compared with a precision 2 parts in 10^{11}[12] and times can be compared with a precision of 3 parts in 10^{16}[13]. While the precision of the kilogramme is adequate for most industrial and scientific purposes, it has been a concern of physicists for some time that the primary standard is not based on a universal property of material, but rather is based on a laboratory artefact – a platinum-iridium alloy cylinder. Secondary standards exist around the world at other laboratories (e.g. NIST, Boulder, USA; Greenwich as well as others), and they are periodically checked ('weighed') against the prototype. Hypothetically, you could imagine the situation in which the primary standard is destroyed, e.g. USA declaring nuclear war on France! After the destruction, the kilogramme secondary standards from the remaining labs would be averaged to give a new primary standard. More seriously there is a risk of the kilogramme prototype drifting in mass over the decades due to the accumulation of dirt. Comparisons between the secondary standards and the international prototype have already revealed an apparent upwards drift relative to the prototype[14]. Whether the prototype has been losing mass, or the secondary standards gaining it, is unclear. More likely, both the prototype and the secondary standards are gaining mass at different rates, although they do undergo periodic cleaning to minimise the impact of this. As a result of these uncertainties, one of the goals of metrology is to replace the kilogramme by a universal standard like the metre and second. To do this requires reproducing a physical process depending on mass, or equivalently, power, which can duplicate the same or better precision than current comparisons of test masses with the prototype.

Physicists believe that this goal will shortly be achieved, as described in Chapter 17. However, changes cannot be adopted until experimental precision is improved, perhaps by a further factor of two, to make the transition from the international prototype kilogramme worthwhile. This is an exciting period in the definition of our unit systems!

2.6 The Fundamental Physical Constants

One of the goals of Physics is to make improved measurements of the physical constants. These represent quantities which can be considered as the fundamental quantities which determine the properties of the universe, and include the speed of light (c), the strength of gravity (G), the charge on the electron (e) etc. For certain key measurements, if they can be made to high enough precision, they can be used to redefine our unit system. Over a period of decades for example, measurements of c became more accurate, attaining the value 299,792,458 m s^{-1}. By this time, no further improvement could be made since the metre itself was not defined to such precision. The switch was then made to fix c and to define the metre by timing a laser beam. The method provided higher reproducibility than the old metre standard, based on a fixed number of wavelengths of light emitted by krypton gas. Should our physics understanding change in future, and c no longer be thought of as fixed for all observers, then the primary standard would need to be changed again, although the chance of this happening would seem to be quite slim.

Measuring the World, by John Austin

In the following list of constants, the precision of each measurement is specified. This is the expected agreement that can be reached by independent measurements of the same quantity, and varies substantially depending on the constant.

The Fundamental Physical Constants: abbreviated list based on the 2010 CODATA adjustment[15].

Measurement	Symbol	Value	Precision	Chapter
Velocity of light in a vacuum	c	299792458 m s^{-1}	Exact	5
Charge on the electron	e	1.60217657 × 10^{-19} C	2 in 10^8	11
Gravitational constant	G	6.6738× 10^{-11} N kg^{-2} m^{-2}	1 in 10^4	7
Mass of proton	m_p	1.6726218× 10^{-27} kg	4 in 10^8	7
Avogadro constant	N_A	6.0221413× 10^{23} mol^{-1}	4 in 10^8	12
Mass of electron	m_e	9.1093829× 10^{-31} kg	4 in 10^8	7
Planck Constant	h	6.6260696× 10^{-34} J s	4 in 10^8	9
Boltzmann Constant	k	1.380649× 10^{-23} J K^{-1}	9 in 10^7	9
Molar gas constant	R	8.314462 J mol^{-1}K^{-1}	9 in 10^7	9
Stefan-Boltzmann constant	σ	5.67037× 10^{-8} W m^{-2} K^{-4}	4 in 10^6	14
Mass of Neutron	m_n	1.67492735× 10^{-27} kg	4 in 10^8	7

Measuring the World, by John Austin

The poorest-known constant is G, which is known only to 1 part in ten thousand. It is needed to determine that gravitational force, F, between two objects of mass m_1 and m_2 : $F = Gm_1m_2/r^2$ where r is the distance between their centres of mass. Compared with other forces such as electromagnetism, gravity is a weak force, so G is poorly known. Research is continually under way to improve the measurements of the physical constants, i.e. to reduce the uncertainties quoted. The values in the table shown are based on the 2010 Committee on Data for Science and Technology recommendations[15]. Significant transitions can occur occasionally when measurements can be done to higher precision than the primary standards allow. After a period of deliberation at the highest scientific levels, the SI system has then changed standards, with the metre described above as the best example. In this case, improvements in laser technology enabled distances to be more accurately determined by timing a laser beam, so by specifying the velocity of light, distances can in principle be measured to the same (high) precision as time. Practical matters dictate a lower precision in practice for realising the metre standard, but the precision is still higher than using the previous wavelength standard. Physicists' strategy for the unit system is first to translate units to universal standards and to make those standards more and more reproducible by improving measurements of the physical constants. When the primary standard is redefined, the definitions no longer contain round numbers. For example, the second isn't 10^{10} periods of the Caesium-133 transition (see Chapter 3), and the metre isn't the distance travelled by light in $1/3 \times 10^8$ seconds (Chapter 5), but they are close (differences of 8% and 0.7% respectively). The reason for this is that these awkward definitions fit the previous ones without needing to change the unit sizes by a detectable amount. The units could be adjusted, but this would cause considerable impacts on society. Changes to the second would be particularly inconvenient as it would no longer fit with the calendar. As it is, the calendar is not perfect (Chapter 3) but an 8% adjustment would be difficult to work into a new calendar. Furthermore, in future we might discover new experimental techniques which improve even more the reproducibility of the units. The power of these methods is that the primary standards can be improved according to the demands of industry and technology without the front end user needing to be aware of the process. For example, GPS tracking during the last decade has put pressure on measurement precision and the primary standards to triangulate exact positions (see Chapter 3).

2.7 Scale Analysis and Alternative Unit Systems

The fundamental physical constants give possible pointers to the natural unit system of the universe. Using a well-known procedure in science known as scale analysis we can explore this approach. For example if we were to represent measurements in terms of the three fundamental quantities mass (M), length (L) and time (T), speed for example has dimensions of L/T. The important point of scale analysis is that the dimensions are mutually exclusive and a practical use of scale analysis is to check the units of a measurement or combination of measurements. Scale analysis can also be used to provide insight into physical processes. If we take the three constants c, G and h (in practice we use $\hbar = h/2\pi$ as it is more fundamental in quantum mechanics), we can rearrange our units such that c, G and $\hbar$ are exactly 1. In this example, c has units LT^{-1}, G has units $[force]L^2M^{-2}$. In turn a force is a *mass x acceleration*, which is MLT^{-2}. Putting this together we see that G has units of $M^{-1}L^3T^{-2}$. As seen in the above table, $\hbar$ has units of Js or in the current notation ML^2T^{-1}. Combining all this together we get

$$c = LT^{-1} \qquad G = M^{-1}L^3T^{-2} \qquad \hbar = ML^2T^{-1}$$

and we find that

Measuring the World, by John Austin

$$G\hbar/c^3 = L^2 \qquad c\hbar/G = M^2 \qquad G\hbar/c^5 = T^2$$

in other words, our new scales can be written

$$L = (G\hbar/c^3)^{1/2} \qquad M = (c\hbar/G)^{1/2} \qquad T = (G\hbar/c^5)^{1/2}$$

These are known Planck scale and Planck units[16] and this is the spatial and temporal scales over which the forces have a similar strength, although the exact mechanism for this 'unification' is unknown. In cosmological terms, these scales were important shortly after the big bang. It also represents the scale on which our current laws of physics break down. The Planck scale is currently an important area of research for theoretical physicists.

The Planck units could in principle be used to communicate information about scale sizes on our planet to a distant civilisation[17]. In practice, the units are very different in size to that experienced in the universe at the current time, a problem which has troubled astronomers for over a century. Substituting in known values for c, G and $\hbar$, we find that

$$L = 1.616 \times 10^{-35} \text{ m} \qquad M = 2.177 \times 10^{-8} \text{ kg} \qquad T = 5.391 \times 10^{-44} \text{ s}$$

These scales are extremely small, and most relevant to the subatomic scale. Nonetheless, the mass scale is much larger in size than the others by several orders of magnitude and it reflects the fact that gravity, which determines the mass scale relative to the others, is an extremely weak force compared with other forces such as electromagnetism. The possible implication is that soon after the big bang the forces were much more comparable, and one of the enduring goals of physics is to 'unify' all the forces by explaining them all with the same equations. A similar procedure to that above can be used to determine temperature and electrical charge scales[18].

Although this procedure leads to unit scales that are vastly different to every day experience, and therefore impractical, a modification of the approach can be useful. Instead of setting c, G and $\hbar$ equal to 1 in the example above, we can instead set them to any constant that we please. In particular, we describe in

Chapter 17 how this procedure has been used to redesign the SI system by specifying an appropriate set of fundamental constants.

2.8 Practical Scale Analysis

A more concrete use of scale analysis is to estimate the size of a second object given limited information about it, together with detailed information about the first object. For example, if we know the mass m and length l of an object, how will this scale up or down to a larger or smaller object? To first order we can assume the density of the objects are the same, and that the objects are exactly the same shape. Here, the volume is proportional to the length cubed so for constant density $m = al^3$ to first order where a is some constant. The idea has some flexibility and generality. For example, what is the approximate mass of a mouse of length 7.5-10 cm (excluding tail). You can guess by assuming that the human and mouse shapes (excluding the tail) are the same. If subscript h is for human and m for mouse we find that $m_m/m_h = l_m^3/l_h^3$. Putting in my human values $m_h = 65$ kg, $l_h = 1.72$ m, and taking the mid value for the length of the mouse, $l_m = 8.75$ cm, then $m_m = 65 \times 0.0875^3/1.72^3 = 65 \times 6.7 \times 10^{-4}/1.72^3 = 8.6 \times 10^{-3}$ or 9 g approximately. This is a bit on the small side: a typical mouse is about 10 - 25 g[19], about twice the predicted value, but the method is only intended to be approximate.

2.9 Computing Units

The metric system has been adopted by the computer industry with minor modifications. Computers of course work with bits, or powers of 2. It is convenient that $2^{10} = 1024 \approx 1000$. So in computing terminology, kb (kilobits) is usually used to denote 2^{10} bits, and kB is usually used for 2^{10} Bytes, where 1 Byte = 8 bits, a convenient number of bits to allow distinct characters (A,B,C,....) to be represented. The above multipliers have been extended upwards as computing power has developed. Thus 1 MB = 1024 kB etc. In large computing establishments, GB, TB, PB and EB are recognised.

While the above notation is often used, it is not recommended by the SI on the principle that it is ambiguous because of the confusion between *k* meaning 1000 or 1024. Instead, the following precise notation is recommended[20]: kibi (ki) = 2^{10}; mebi (Mi) = 2^{20}; gibi (Gi) = 2^{30}; tebi (Ti) = 2^{40}; pebi (Pi) = 2^{50}; exbi (Ei) = 2^{60}. Therefore a kibibyte would be written kiB etc. Clearly, major computing installations need to get hold of this idea and start using it regularly so that the terms can become familiar.

References

[1] British Imperial System, Encyclopedia Britannica, http://www.britannica.com/EBchecked/topic/80231/British-Imperial-System, accessed 13 February 2014.
[2] Introduction to the Metric System, Wikipedia, http://en.wikipedia.org/wiki/Introduction_to_the_metric_system, accessed 13 February 2014.
[3] Imperial Yard Shrinking, *The Sydney Morning Herald* (NSW, Australia), 8 June, 1950, p. 3., http://nla.gov.au/nla.news-article18159945, accessed February 13, 2014.
[4] Girard, G., 1994: The third periodic verification of national prototypes of the kilogramme (1988-1992), Metrologia, 31, 317-336, doi:10.1088/0026-1394/31/4/007
[5] McCarthy, D.D & Seidelmann, P.K., 2009: *TIME: From Earth Rotation to Atomic Physics.* Weinheim: Wiley-VCH, pp. 88–89.
[6] International Yard and Pound, Wikipedia, http://en.wikipedia.org/wiki/International_yard_and_pound, accessed February 13, 2014.
[7] The International System of Units (SI) – and the "New SI", Bureau International des Poids et Mesures, Sèvres, Paris, France,

Measuring the World, by John Austin

http://www.bipm.org/en/si/, accessed February 13, 2014.
[8] Thousandth of an inch, Wikipedia, http://en.wikipedia.org/wiki/Thousandth_of_an_inch, accessed 14 February 2014.
[9] Drax – facts and figures -power station, http://www.draxpower.presscentre.com/Facts-and-figures/Power-station-b89.aspx, accessed 14 February 2014.
[10] Edwin Cartlidge, Breaking News: Error undoes faster-than-light neutrino results, Science Magazine, 22 February 2012, http://news.sciencemag.org/2012/02/breaking-news-error-undoes-faster-light-neutrino-results, accessed 15 February, 2014.
[11] Barat, P, Fang, H and Goyon, C., 2012: Calibrations at BIPM using the present definition of the kilogram, CCM Workshop on the mise en pratique of the new definition of the kilogram, 21-22 November 2012, Sèvres, France, http://www.bipm.org/ws/CCM/MeP_2012/Allowed/November_2012/12a_6.1_BIPMMassCalibrations_WS_21nov_final.pdf, accessed 15 February 2014.
[12] Iodine ($\lambda \approx 633$ *nm*), 2003: Mise en Pratique, Bureau International des Poids et Mesures, Sèvres, France, http://www.bipm.org/utils/common/pdf/mep/M-e-P_l2_633.pdf, accessed 15 February, 2014.
[13] NIST-F1 Cesium fountain clock, National Institute of Standards and Technology, Physical Measurement Laboratory (USA), 4 February 2013, http://www.nist.gov/pml/div688/grp50/primary-frequency-standards.cfm, accessed 15 February 2014.
[14] The BIPM watt balance → towards a definition of the kilogram, Bureau International des Poids et Mesures, Sèvres, France, http://www.bipm.org/en/scientific/elec/watt_balance/, accessed 15 February 2014.
[15] Mohr, P. et al., 2012: CODATA recommended values of the fundamental physical constants: 2010, Reviews of Modern Physics, Vol. 84, p.1528-1605.
[16] Addler, R.J., 2010: Six easy roads to the Planck scale, http://arxiv.org/pdf/1001.1205v1.pdf, accessed 17 February 2014.
[17] Busch, M.W., and Reddick, R.M., 2010: Testing SETI message designs, Astrobiological Science Conference 2010,

http://www.lpi.usra.edu/meetings/abscicon2010/pdf/5070.pdf, accessed 17 February 2014.
[18] Planck units, Wikipedia, 28 January 2014, http://en.wikipedia.org/wiki/Planck_units, accessed 17 February 2014.
[19] House mouse, Wikipedia, 12 February 2014, http://en.wikipedia.org/wiki/House_mouse, accessed 17 February 2014.
[20] Prefixes for binary multiples, 2014: International Electrotechnical Commission, http://www.iec.ch/si/binary.htm, accessed 17 February 2014.

Measuring the World, by John Austin

3. Time

Time = the duration elapsed between events

3.1 Introduction

Time is unlike all other measurements, and is arguably the most fundamental. Unlike distance for example, time always increases as immortalised by the Omar Khaiyam in his Rubaiyat from several millennia ago:
> The moving finger writes; and having writ
> Moves on: nor all thy piety and wit
> Shall lure it back to cancel half a line,
> Nor all thy tears wash out a word of it

Although we all have an intuitive understanding of time, when pressed to define it, no one single response is forthcoming. For example Carroll[1], arrives at the following three broad ideas: 1. It is a label to measure moments in the universe. 2. It is a duration and is that which is measured with a clock. 3. Time is a medium through which we move from past to present, toward the future.

In principle, time can be understood through the laws of thermodynamics, which govern all physical processes in the universe. These laws, established over centuries of physical measurements and theories are as follows:
1. The combined mass and energy of a closed system is conserved.

2. The entropy of an isolated system remains constant, or increases with time.
3. As the temperature of a system is lowered, there is a minimum (absolute zero) for which the entropy is also a minimum.

Mass and energy are discussed in Chapters 7 and 11. Temperature is discussed in Chapter 8. This leaves "entropy" in law 2 which could be used to understand time better. Entropy is a measure of the disorder in a system, so the 2nd law indicates that the disorder in a system increases with time[2]. For example if you add a spoon-full of sugar to a cup of coffee and stir it in, the mixture is in a more disordered state than the coffee and sugar separately. Once mixed (high entropy), the sugar can't be unmixed again (low entropy). Entropy is quantifiable using the relation $S = klogW$. S is entropy and k is Boltzmann's constant, which has already appeared in §2.6. W is the "number of particular microscopic arrangements of atoms that appear indistinguishable from a microscopic perspective". Entropy is an important quantity for physicists but is tangential to our main topics here. Nonetheless, it can be born in mind that a clearer concept of time may be used by invoking entropy, and at the same time, the 2nd law of thermodynamics tells us why we can't go backwards in time. Time, then can be thought of as "that which clocks measure".

Traditionally, it was thought that time was absolute, the same for all observers. However, one of the main conclusions of Special Relativity (SR) is that with accurate clocks, different observers can have clocks which run at different rates depending on the velocities of the observers[3]. The clock rate can also change in the near massive objects such as a star, as in the more complete theory of General relativity (GR). Ordinarily, the time difference between different observers is not noticeable, until the velocity approaches the speed of light. A further conclusion from SR is that time travel is possible, but only into the future. Once in the future, you can't come back. For example, to travel into the

Measuring the World, by John Austin

future, you need to get into a spaceship and continue at very high speed. You then slow down, turn around and accelerate to high speed again. Depending on your trajectory, the journey might take 1 year as far as your own clocks are concerned (and your body will have aged 1 year), but 10 years might have elapsed on Earth during your absence. Don't get too excited about these prospects, though. Ignoring the slowing down period and time taken to turn around, you still need to travel at 99.5% of the speed of light to achieve a factor of 10 between the two clocks in the above scenario. By comparison, the fastest man-made object, the Voyager 2 craft, is now leaving the solar system at a speed relative to the sun of `only' 17 km s^{-1} or 5.7x10^{-5} c [4].

People have been measuring time for as long as recorded history. For times longer than a day, calendars have been used to mark the changing of the seasons and over the centuries, the calendars have become more accurate once the length of the year was accurately known (§ 3.2). Our current calendar is described in § 3.3, but this is by no means the only possible calendar, and a simpler and better one is described in § 3.4. For short periods, clocks have been invented. These were often based on the Earth, but in time the precision needed increased and hence clocks were invented, based on more stable physical processes (§ 3.5). Timing methods have improved enormously over recent decades, and modern atomic clocks are capable of very high precision, better than 1 part in 10^{15}, as described in § 3.6, although high quality pendulum clocks are still being constructed (§ 3.7). The second, described in § 3.8 is now the linchpin of the SI system of units, since with the specified velocity of light, it now supports the unit of length (Chapter 5). The second as now defined is more stable than the old Earth standard, and the slowing of the Earth's rotation over long duration has necessitated the occasional adjustment of our calendars, as described in § 3.9. Examples of timing in sport are discussed in § 3.10.

3.2 History of the Calendar

Ancient civilisations have provided historical documentation of the calendars that they have used. Records show that the Chinese calendar began in 2637 BC. It had cycles of 60 years with 12 sub cycles. The calendar receives a lot of publicity near the time of the Chinese new year, which is usually announced with colourful celebrations[5]. It starts at the 2nd moon after the Northern winter solstice. Dates are adjusted empirically with the new year starting at any day within the range 21 January to 20 February (in our modern calendar) depending of course on the phase of the moon inn the previous year. During the 12 year cycles, the years are named after the animals (in sequence) rat, ox, tiger, hare, dragon, snake, horse, sheep, monkey, rooster, dog and pig. 2014 will be the year of the horse. Similar to other astrological superstitions, years of particular animals are imbued with certain properties. There are 12 months in the calendar, with alternating 29 and 30 days, staying in phase with the moon, which has an orbital period around the Earth of 29.5 days.

The Islamic calendar[6] also has 12 months alternating between 29 and 30 days. This gives a total of 354 days per year, so the new year moves backwards through the seasons over a period 365.24/11.24 = 32.5 years. The calendar is put almost back on track with the seasons with 30 year cycles. It is difficult to believe that this impractical calendar would still be used anywhere.

Once the Earth's seasonal cycle was properly recognised, time standards were established more precisely. In 1580, essentially long before Europeans started to build their empires, the Gregorian calendar as it is now known, was invented by Pope Gregory XIII. It provided a fairly accurate representation of the annual movement around the sun, but Britain continued to use the old Julian Calendar supplied by the Romans. By the 18th century, the Julian calendar had gradually drifted away from an accurate calendar, by about 11 days, as the treatment of leap years was not correct. Finally, in 1752 Britain adopted the Gregorian calendar amidst public outrage. Because of the 11 day adjustment[7], many citizens protested against 'losing' 11 days of their lives[8]. It is easy for us to mock this behaviour in these more sophisticated times, but is this much different to the 50% of the US population that currently deny the influence of humans on Earth's climate?

The modern (Gregorian) calendar[9] is based on Christian theology. The year of Christ's birth is attached to 1 AD, and the year before is 1 BC, although the precise details are unknown. This means, incidentally that the start of the millennium was 2000 years after the start of the calendar, or 1/1/2001. Like most people today, though, I find the year 2000 more emotive and consider it an oversight on Pope Gregory's part, not to have a year 0. I just wish we could agree to start counting decades, centuries and millennia from 1st of January of years ending in 0's and just be done with it. The discussion seems to come up needlessly (with no resolution) every decade with varying passion, depending on the number of 0's in the year. The AD/BC terminology has now been changed to avoid offending non-Christian religious individuals that are easily offended, and it is now common to use CE (Current Era) in place of AD, and BCE (Before the Current Era) in place of BC. Of course if the Muslims had known the number of days in the year, and constructed a more sensible calendar, we could now be measuring years after Mohammed, and many of us might not care a jot. Despite the occasional use of these more inaccurate calendars, the Gregorian calendar seems to be almost globally adopted.

3.3 The Gregorian Calendar

The key objective of any calendar is to keep the

seasons at fixed dates. This implies the need to know the length of the year to reasonably high precision. Modern measurements put the length of the year at $365.24219647 - 6.24\times10^{-6}T$[10] days, where T is the time in centuries since 1 January 1900. One day is 24 hours exactly (86400 seconds). Each day consists of the 23 hours 56 minutes 4.09 seconds rotation of the earth on its axis (sidereal day) plus an extra period just under 4 minutes which maintains the sun at the highest point in the sky at local noon. So, with a 24 hour clock, the stars rise 4 minutes earlier each night. The small fraction of 0.242 days requires the insertion of an extra day ("leap day") periodically to keep the seasons in place. The Gregorian calendar has a simple method to do this and which is so engrained into our culture that we scarcely think about it.

The standard calendar with 28 days in February has 365 days in total. After 750 years, the difference from the true (astronomically correct) calendar is 365.242x750 – 365x750 = 181 days. This means that the seasons would be completely reversed compared to the current time. Over a period of 1000 years, we would need to add an extra 242 days to keep the seasons aligned at their current dates. If we were to add one new day every 4th year, for example 2000, 2004, 2008, 2012,.... then we would have 250 days extra which is too many. This was the fault of the old Julian calendar. Suppose that we exclude all the century years, then we would have 240 leap days, which is only 2 too few. Finally, suppose that we accept as leap days those that occur every 400 years. For example, 2000 would be a leap year, but 2100 would not be. Then we would have 242 or 243 extra days in our calendar, very close to the desired 242. Each of the extra days is known as a "leap day" and a little inconsistently, the year it occurs in is known as a "leap year". This procedure defines the length of the calendar as 365.2425 days, close to the actual value of 365.242. Since the sun is moving through the galaxy, the length of the year relative to the fixed stars is slightly different at 365d 6h 9min 9.54s (365.25636 days). These exquisitely high precision (3 parts in 10^{11} for the length of the year) measurements mean that the year is slowing at the rate of $6.24 \times 10^{-6} \times 86400 = 0.539$ s per century!

While these high precision measurements are interesting to know about, the issue of leap years has been known to cause commercial problems. Failure to understand the century rule properly, led to the publication of some diaries in 2000 with 29th February missing! After all these years, I can no longer find the report about them, but I'm sure it's not a figment of my imagination. No doubt some mistakes will be made in 2100 as well, but that is still some decades away.

While there is agreement on the use of a calendar, there are still differences in the representation of the date. For example, in the UK, and many British Commonwealth countries the date is written as *dd/mm/yy* where *dd* is the day number, *mm* is the month number and *yy* the two digit year number. On the continent of Europe and many other countries the date is indicated by *yy/mm/dd* while in the USA the date is represented by *mm/dd/yy*. The latter is quite illogical and I would like to see it disappear. In my 10 years living in the USA, I was more confused by that than any other notation that I saw on a regular basis. Even in the USA, the practice is not adhered to, as the immigration forms that you need to complete on entering the country are in the "international order" *yy/mm/dd*.

3.4 The World Calendar

Gregorian calendar, although standard, appears rather quaint with its 28, 30 or 31 day months. Amongst those calendars which have been proposed is the 'World Calendar' [11](diagram), consisting of 4 quarters, plus 1 or 2 `worldsdays'.

Measuring the World, by John Austin

January						
S	M	T	W	T	F	S
1	2	3	4	5	6	7
8	9	10	11	12	13	14
15	16	17	18	19	20	21
22	23	24	25	26	27	28
29	30	31				

February						
S	M	T	W	T	F	S
			1	2	3	4
5	6	7	8	9	10	11
12	13	14	15	16	17	18
19	20	21	22	23	24	25
26	27	28	29	30		

March						
S	M	T	W	T	F	S
					1	2
3	4	5	6	7	8	9
10	11	12	13	14	15	16
17	18	19	20	21	22	23
24	25	26	27	28	29	30

April						
S	M	T	W	T	F	S
1	2	3	4	5	6	7
8	9	10	11	12	13	14
15	16	17	18	19	20	21
22	23	24	25	26	27	28
29	30	31				

May						
S	M	T	W	T	F	S
			1	2	3	4
5	6	7	8	9	10	11
12	13	14	15	16	17	18
19	20	21	22	23	24	25
26	27	28	29	30		

June						
S	M	T	W	T	F	S
					1	2
3	4	5	6	7	8	9
10	11	12	13	14	15	16
17	18	19	20	21	22	23
24	25	26	27	28	29	30 W

July						
S	M	T	W	T	F	S
1	2	3	4	5	6	7
8	9	10	11	12	13	14
15	16	17	18	19	20	21
22	23	24	25	26	27	28
29	30	31				

August						
S	M	T	W	T	F	S
			1	2	3	4
5	6	7	8	9	10	11
12	13	14	15	16	17	18
19	20	21	22	23	24	25
26	27	28	29	30		

September						
S	M	T	W	T	F	S
					1	2
3	4	5	6	7	8	9
10	11	12	13	14	15	16
17	18	19	20	21	22	23
24	25	26	27	28	29	30

October						
S	M	T	W	T	F	S
1	2	3	4	5	6	7
8	9	10	11	12	13	14
15	16	17	18	19	20	21
22	23	24	25	26	27	28
29	30	31				

November						
S	M	T	W	T	F	S
			1	2	3	4
5	6	7	8	9	10	11
12	13	14	15	16	17	18
19	20	21	22	23	24	25
26	27	28	29	30		

December						
S	M	T	W	T	F	S
					1	2
3	4	5	6	7	8	9
10	11	12	13	14	15	16
17	18	19	20	21	22	23
24	25	26	27	28	29	30 W

Each month would begin on the same day every year and the 1 or 2 year days would not have a specific 'day of the week'. The calendar is designed with 30 and 31 day months, so that each 3 month period is exactly 13 weeks, and starts always on a Sunday. A worldsday (W in the diagram) is inserted after the end of December, and on leap years an additional worldsday is inserted after the end of June. Such a calendar is undoubtedly superior to the confusing Gregorian calendar, but as no country is currently using it, in conjunction with the usual religious objections, means that it probably has little prospect of taking off. The question is why Pope Gregory XIII didn't design such a calendar in the first place.

The UN considered implementing the calendar in March 1955, but it was opposed by the US on religious grounds. Of course if religious people always had there way, we would still be using sundials, and would still think the sun revolves around the Earth.

3.5 A History of Early Clocks

Since time immemorial, it has been recognised that the movement of the sun can be used as a clock during the daytime to divide time into units smaller than a day. Modern 'sundials' as they are known, are frequently used as decorative pieces in gardens. Sundials usually consist of a horizontal platform (occasionally a vertical platform) in the centre of which is mounted a 'gnomon' which produces a shadow on the platform. The edge of the shadow indicates the time. The upper edge of the gnomon slants upwards from the dial face at an angle equal to the latitude. The sundial gives the local time but needs to be corrected for changes in the Earth's orbit during the year. The correction is known as the 'equation of time' and varies from just a few to almost 20 minutes [12]. The local time is displaced from the time on our clocks ('civil time') according to the longitude. For example, at the Greenwich Meridian (0° W), a sundial, after correction for the equation of time , would give the correct local time of 12.00 pm. At 15° W, at the same instant, the sundial would indicate a time of 11.00am. All-in-all, a sundial is a somewhat imprecise device for giving the time, and of course doesn't work at night or in cloudy weather!

In about 1500 BC, the Egyptians used water clocks[13], although they may have been in use in China as early as 4000 BC[14]. These consisted of a container with scale markings and as water ran out of the container, the water left in the container marked the time. The Romans also used water clocks. The flow of water depends on its viscosity (fluid friction) which varies substantially with temperature. Consequently, an accurate clock would need careful control of temperature and there was no evidence that this was done in ancient times. Water clocks must therefore have been quite unreliable, varying by half an hour to an hour per day. Sand clocks worked on the same principle: a constant flow of sand is converted to an elapsed time. Typically known as hourglasses, they were not routinely used until the 14th century[15]. Sand clocks are still used today in the form of decorative egg timers!

Far higher accuracy was realisable with pendulum clocks which were invented in 1656 by Cristiaan Huygens, based on principles discovered by Galileo Galilei[16]. In particular, the period of the pendulum is independent of the amplitude of the pendulum swing providing the amplitude of the swing is small – 5° or so. It became possible to construct clocks which were accurate to a few seconds per day, and by 1930, pendulum clocks were the most precise available, capable of an accuracy of 10 *ms* per day (1 part in 10^7)[16]. Pendulum clocks were in due course replaced by electronic clocks for high precision work. In addition to measuring time, clocks provided the unexpected solution to the problem of measuring longitude at sea, as described in Chapter 5.

3.6 A History of the Atomic Clock

The idea behind the atomic clock was first suggested in 1879 by Lord Kelvin[17] but was not actually developed until the 1930s. By 1945 atomic beam magnetic resonance had been perfected to allow a clock to be developed[18]. The first accurate atomic clock, based on a transition frequency in the Caesium-133 atom, was built in 1955[19] and in 1958 this technique was used as a standard for the SI second (§ 3.8).

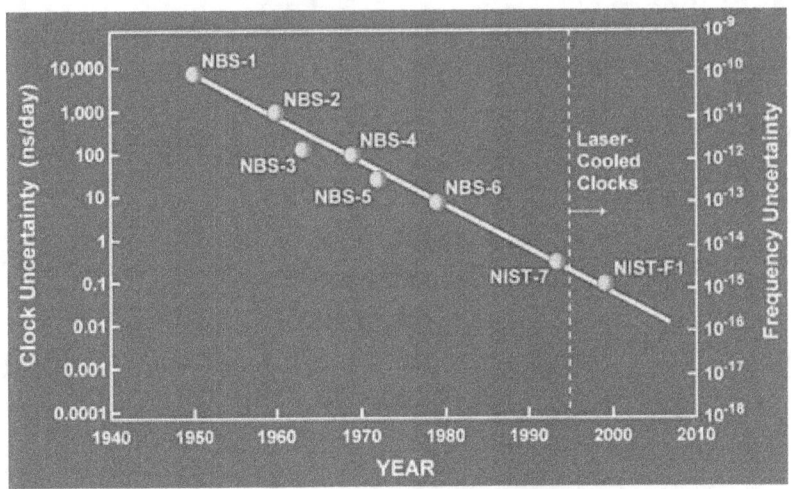

Clock uncertainty in *ns* per day at NIST[20] since 1950.

Current atomic clocks are also based on caesium atoms and timing intervals are determined by measuring the number of oscillations at the fixed frequency of the electromagnetic radiation emitted. The clock precision of 1 second in 100 million years has been reached (3 parts in 10^{16}) by the NIST-F1 fountain clock[20], the best of any of the primary units. The above figure, published by NIST, shows the clock uncertainty in ns per day since 1950. In that year, the clock uncertainty amounted to some 10,000 ns per day, or 1 part in 10^{10}, but timing technology has evolved over the decades, stepping throught the different named clocks until an uncertainty of 0.1 ns per day was attained in about the year 2000. Progress has been steady during the last 50 years with the clock uncertainty reducing by about a factor of 10 each decade. Since NIST-F1 has been in operation, clock uncertainty has been reduced by annother factor of 10, and current uncertainty levels are approaching 1 part in 10^{16}, or an uncertainty of only 45 seconds over the lifetime of the universe, 4.34×10^{17} s.

3.7 Modern Pendulum Clocks

The Longnow 10,000 year clock, Prototype 1[21], currently on loan to the Science Museum of London. Photo by Rolfe Horn

Conventional clock building is still alive and well in the age of the atomic clock, primarily for demonstration purposes. A mechanical clock is being built to be housed inside a 150 m deep shaft in west Texas[21], with the objectives of keeping the clock running correctly for 10,000 years. It will have a 3 m pendulum weighing 110 kg. Prototype 1 was completed on 31 December 1999, and is displayed in the London Science Museum. It has a 2.4 m pendulum and the clock can be adjusted for the variations in the length of the day. The time will be displayed when explicitly requested by a visitor. The estimate is for the clock to remain accurate to the nearest second for its projected life time of 10,000 years, giving a precision of about 3 parts in 10^{11}. This is not as good as the best atomic clocks, but it will provide visitors with an understanding of the major engineering problems in constructing clocks.

3.8 The Second

The second is the unit of time used in all unit systems. It was originally defined as a fraction 1/86400 of a day, based on the mean day length. This is the length of the solar day: the time between the high point of the sun on one day to the high point of the sun on the next day. The total number of seconds per day that we still use is based on the ancient Babylonian mathematics of about 2000 BC[22], consisting of 60 seconds per minute and 60 minutes per hour. As with other units, especially the metre (Chapter 5), the idea of modern unit systems was to have an Earth-based artefact to provide the primary standard and to use clocks to provide a secondary standard, with units broken into smaller, more manageable sizes.

Problems arose when high precision time measurements were needed. It was found that the Earth's rotation rate was not constant, but the length of the day varied by up to 1 ms[23], i.e. 1 part in 10^8. So this limited the precision at which times, based on the mean solar day, could be compared. The changes in rotation rate arise from the forces exerted by the atmosphere and ocean on the solid Earth[24]. Major storm systems are the main atmospheric contributors. Without external forces, the angular momentum (or equivalently the rotation rate) of the Earth + Atmosphere + Ocean is constant. This is the rotational equivalent of the conservation of linear momentum (see Chapter 9). Forces between the solid Earth and the atmosphere can change the rotation rate of the atmosphere and the solid Earth so that net angular momentum is conserved. For example, storms which decrease angular momentum will increase the Earth's rotation rate and hence decrease the length of the day. Compared with the mass of the Earth, the atmospheric mass is much smaller and so the effects of the atmosphere on the Earth are rather small. Nonetheless, they are detectable[23, 24].

The demands of higher precision time measurements stimulated and satisfied by the invention of the atomic clock, resulted in a change in the definition of the second. This started the process of changing the units to universal standards, a quest that has very nearly reached its conclusion (see Chapter 17). In 1967, the new SI definition of the second was introduced as[25]:

"*The duration of 9 192 631 770 periods of the electron transition between two hyperfine ground-state levels of Caesium-133.*"

This is a somewhat technical definition and demands some explanation. The first point is that the relevant observation is the number of waves of radiation in the microwave part of the spectrum (Chapter 5), need to be counted to determine time. The length of the second is unchanged relative to previous definitions, and to match them exactly requires the specification of an awkward number of periods to be counted: not 9 billion, not 10 billion, but 9.192.... billion. The new definition allows times to be compared to a precision of at least 1 part in 10 billion, i.e., 1 part in 10^{10}, compared with 1 part in 10^8 if the rotation of the Earth is used. What the above definition means is that starting from Caesium-133 gas the equilibrium (ground state) is perturbed by radiation and it will then emit like a laser in a very precise wavelength with the above frequency. The wavelength, λ is c/f where c is the speed of light and f is the frequency. So, $\lambda \approx 3 \times 10^8 / 9 \times 10^9 \approx 0.03$ m, or about 3 cm wavelength. It is a low energy transition, but highly reproducible. Caesium is a soft metallic element[26] which is present in small concentrations in the form of its carbonate and chloride salts dissolved in mineral waters and sea water. Its atomic number is 55 (55 protons in the nucleus). It melts at 28.4 °C, and boils at 641 °C. The metal is usually made from the chloride salt, CsCl by high temperature reduction. Caesium-133 is a particular isotope of Caesium. Specific elements have fixed number of protons in the nucleus, and different isotopes are marked by the different number of neutrons in the nucleus. The number of protons and neutrons in the caesium nucleus is 133, and the number of protons (atomic number) is 55, so the number of neutrons is 133 − 55 = 78. Caesium is an alkali metal (like sodium and potassium) which can be purified to an exceptional high degree. The high purification makes caesium-133 an ideal standard. As noted in chapter 2, time can be measured to the highest precision of any quantity in physics. For this reason, it is used with the specified velocity of light, to define the metre (and hence the foot) (see Chapter 5).

3.9 Leap Seconds

The changes in the Earth's rotation rate due to the atmosphere and ocean are essentially random over long time scales. Some weeks are very stormy, and the length of day decreases, while other weeks are less stormy, and the length of day increases. Nonetheless, over long periods of time (centuries and millennia) the effect of the atmosphere and ocean forces is to slow the rotation rate, with the excess energy dissipated as heat. The moon also provides a drag on the Earth, slowing the rotation further.

To catch up with solar time requires an additional rotation of the Earth in a complete year. In other words, the Earth rotates 366.242 days in a year, although on Earth it appears as if the sun goes around us 365.242 days, the conventional year length. This gives us a mean sidereal day of 86400x365.242/366.242 seconds, or 23 h 56 min 4.09 s. The exact day length may be slightly longer by a few ms.

Eventually, though, discrepancies with astronomical features mean that these millisecond differences from our mean 24 hour clock accumulate systematically. To keep astronomical features aligned international agreements are in place to add 'leap seconds' to the calendar. The leap second when needed is added at midnight Universal Time, UT, or Greenwich Mean Time (GMT) as it used to be known, on either 30 June or 31 December. This has typically been done once every 1-2 years for the past 40 years. Due to their relative unimportance to most people, they are not heavily publicised. At the time of writing, the last leap second was inserted on 30 June 2012[27]. A leap second will not be added in June 2014[35] and there are no current plans for leap seconds in the immediate future.

The future of the leap second is a topic of debate at the highest technological level[28]. If leap seconds are no longer used, then even after the next 1000 years, our clocks will only have drifted from its original astronomical alignment by about 10-15 minutes, which is not serious enough to concern the public. Scientifically, the main point is that we maintain atomic clocks, so the primary geocentric unit is not important, and adjusting clocks periodically requires organisation and synchronisation. The astronomical community prefers to keep leap seconds to maintain observational programmes more easily. Hence there are advantages and disadvantages in keeping leap seconds. Currently, all clocks worldwide in critical systems need to be synchronised. So leap seconds are coded explicitly in GPS trackers and aircraft control. For example, if GPS trackers were not synchronised, then position errors equal to the velocity of light x 1 second would occur. This distance, 3×10^8 m, is much larger than the size of the Earth, and GPS tracking would fail completely! In air traffic control, a 1 second error would lead to aircraft positioning errors of 300 m, since commercial jets operate at up to 300 m s^{-1}. This would be a serious error, especially during take off and landing. Other systems and organisations such as banks rely on accurate clocks. A decision regarding the continuance of leap seconds has now been postponed by the International Telecommunications Union until 2015[29], but to this writer, they would seem to be unnecessary.

3.10 Timing in Sport

Timing in sport covers two broad areas. The first area covers team events in which the activity (football, hockey etc.) has a set period. This is easily timed with an ordinary watch of some kind, and no special technology is employed. Many other sports use photoelectric timing devices which start and stop as soon as the sports person crosses the appropriate line.

In track athletics, the finishing time is recorded as the time when the chest crosses the line. This is not the same as the light beam first being broken, since an arm or leg might cross the line first. Instead a photo finish camera records a vertical slice of the object, recording 3000[30] slices per second. This enables the correct finishing time to be recorded by inspection of the photo finish photograph. The photograph puts together the thousands of images one vertical slice at a time. For sportsmen crossing the line at a uniform speed, the image looks similar to what you would get with a still camera at a fixed time. To the naked eye, the images from a photo finish camera usually look similar to that of a still camera. However, the camera is a fixed position at multiple times. An athlete for example who had fallen down just short of the finish line and had got up to finish the race would be moving slower than the other performers. The photo finish camera would record this as grotesquely stretched limbs, since it would take so much longer to pass across the slit camera!

Photofinish photograph of the Olympic 100m final, showing Usain Bolt (Jamaica) winning in a time of 9.63s.

Further, the time recorded is the time since the firing of the gun. At international level, for sprinting the starting blocks are wired to measure the force exerted. It is assumed that it is impossible to react faster than 0.100 s. In other words, if a force is recorded less than 0.1 s *after* the gun has fired, it is deemed a false start[31]. As a point of curiosity, the 100 m Olympic record of 9.63 s was set with a reaction time of 0.165 s[32]. So, the actual time for the 100 m was 9.47 s. Bolt's was a relatively slow start, compared with many athletes, who were much closer to the allowable 0.100 s limit. In athletics, then, it is possible to run the

Measuring the World, by John Austin

fastest race, but lose because of a poor reaction time. Another application of physics to the start is that allowance is made for the speed of sound. At international level, for sprinting, each starting position has a loud speaker to convey the starter's commands and the sound of the gun itself. This means that information is conveyed to each athlete simultaneously, ensuring a fair start. For a 100 m race this is not important, as the starter is close to all the athletes, but in 200 m and 400 m races which have staggered starts, this is no longer the case. In these situations, the starter is not necessarily equidistant from all athletes and may be 50 m away from the furthest athlete.

At club level, where speakers are rarely used, I have seen starters standing near the lane 1 athlete, with the lane 8 runner in the 400 m hearing the gun as much as 0.2 s later due to the 330 m s^{-1} speed of sound. Timing is often much simpler, using standard stopwatches and visual inspection for false starts. Accurate timing, which again is most important in the sprints, can be done with care if manual timing only is available. The important point is that the timekeeper must not "anticipate the finish". In other words, you have no warning when the starting gun goes off. The timekeeper starts the watch as soon as he or she sees the smoke or spark, and for a 100 m race, the sound reaches you 0.3 s later. Just like sprinters pushing against their blocks, there is a delay between the command from the brain ("start watch") and its execution. Typically, this delay is 0.2 – 0.3 s. The race takes place, and just as the runner approaches the line you watch carefully, and stop the watch simultaneously with the runner arriving at the line. Wrong! The time will be too fast by 0.2 – 0.3 s because you "anticipated the finish". The correct procedure is to wait for the athlete to finish and immediately issue to yourself the command "stop watch". There will be another delay of 0.2 – 0.3 s and the elapsed time will be correct. In Britain, this was the procedure for the decades prior to automatic timing, which was introduced in the 1970s. Most timekeepers around the world probably anticipated the finish, which necessitated a general increase in manual times by 0.24 s to compare with fully automatic times[33]. Because of the poor times, many top sprinters avoided running in Britain! Eventually, when electronic timekeeping became standard, British timekeepers were vindicated. It is possible, then, to get accurate timings with a manually operated stopwatch (or electronic equivalent in which a person operates the stop and start buttons), although precision of 0.1 s is about the best that can be attained. That is, if international timekeepers had been any good, the 0.24 addition would not have been needed. The same principles of not anticipating the start and finish apply to the timing of any event where precise timing is needed. You can test your own reaction time on the internet here[34]. When I tried it, my average for 5 tries was 313 *ms*, a bit slower than the average of 215.

References

[1] Carroll, 2010: From eternity to here: the quest for the ultimate theory of time, Dutton, Penguin Books Ltd.
[2] Introduction to entropy, Wikipedia, 12 January 2014, http://en.wikipedia.org/wiki/Introduction_to_entropy, accessed 18 February 2014.
[3] Hawking, S., 1988: A brief history of time: from the big bang to black holes, Bantam Dell Publishing Group, 256 pp.
[4] Voyager, The interstellar mission, NASA JPL, http://voyager.jpl.nasa.gov/where/index.html, accessed 20 February 2014.
[5] Chinese New Year, Wikipedia, 14 February 2014, http://en.wikipedia.org/wiki/Chinese_New_Year, accessed 20 February 2014.
[6] Islamic calendar, Wikipedia, 13 February 2014, http://en.wikipedia.org/wiki/Islamic_calendar, accessed 20 February 2014.
[7] The curious history of the Gregorian calendar, Eleven days that never were, Information please, Database 2007,

http://www.infoplease.com/spot/gregorian1.html, accessed 20 February 2014.

[8] Barrow, M., 2013: Old Christmas day and the lost 11 days, Facts of the Day Calendar , http://projectbritain.com/calendar/january/lostdays.html, accessed 20 February 2014.

[9] Gregorian Calendar, Wikipedia, 25 January 2014, http://en.wikipedia.org/wiki/Gregorian_calendar, accessed 20 Februarry 2014.

[10] Meeus, J & Savoie, D. (1992): The history of the tropical year, Journal of the British Astronomical Association, 102(1), 40–42.

[11] The world calendar, Wikipedia, 26 January 2014, http://en.wikipedia.org/wiki/The_World_Calendar, accessed 21 February 2014.

[12] Equation of time, Wikipedia, 28 January 2014, http://en.wikipedia.org/wiki/Equation_of_time, accessed 21 February 2014.

[13] Water clock, Wikipedia, 7 February 2014, http://en.wikipedia.org/wiki/Water_clock, accessed 22 February 2014.

[14] Cowan, Harrison J., 1958: *Time and Its Measurement: From the stone age to the nuclear age*. Ohio: The World Publishing Company.

[15] Hourglass, Wikipedia, 25 December 2013, http://en.wikipedia.org/wiki/Hourglass, accessed 22 February 2014.

[16] Pendulum clock, Wikipedia, 12 February 2014, http://en.wikipedia.org/wiki/Pendulum_clock, accessed 22 February 2014.

[17] Sir William Thomson (Lord Kelvin) and Peter Guthrie Tait, 1879: *Treatise on Natural Philosophy*, 2nd ed. (Cambridge, England: Cambridge University Press), vol. 1.

[18] Isidor I. Rabi, 1945: "Radiofrequency spectroscopy", Richtmyer Memorial Lecture, delivered at Columbia University in New York, New York, on 20 January 1945.

[19] Essen, L., Parry, J. V. L., 1955: "An Atomic Standard of Frequency and Time Interval: A Cæsium Resonator". *Nature,* **176** (4476): 280.

[20] NIST-F1 Cesium fountain clock, National Institute of Standards and Technology, Physical Measurement Laboratory (USA), 4 February 2013, http://www.nist.gov/pml/div688/grp50/primary-frequency-standards.cfm, accessed 22 February 2014.

[21] The 10,000 year clock, http://www.longnow.org/clock, accessed 22 February 2014.
[22] O'Connor, J.J & Robertson, E.F., December 2000: An overview of Babylonian Mathematics, School of Mathematics and Statistics, University of St. Andrews, Scotland, http://www-groups.dcs.st-and.ac.uk/~history/HistTopics/Babylonian_mathematics.html, accessed 23 February 2014.
[23] Fluctuations in the length of day, Wikipedia, 24 December 2013, http://en.wikipedia.org/wiki/Fluctuations_in_the_length_of_day, accessed 23 February 2014.
[24] Hide, R., 1989: "Fluctuations in the Earth's Rotation and the Topography of the Core--Mantle Interface", *Philosophical Transactions of the Royal Society A: Mathematical, Physical and Engineering Sciences,* **328** (1599): 351–363.
[25] Unit of time (second), SI brochure, Section 2.1.1.3, Bureau International des Poids at Mesures, France, http://www.bipm.org/en/si/si_brochure/chapter2/2-1/second.html, accessed 23 February 2014.
[26] Caesium, Wikipedia, 1 February 2014, http://en.wikipedia.org/wiki/Caesium, accessed 23 February 2014.
[27] Leap second, Wikipedia, 19 January, 2014, http://en.wikipedia.org/wiki/Leap_second, accessed 23 February 2014.
[28] Merali, Z., 2011: Time is running out for the leap second, 8 November 2011, Nature, 479, 158, doi:10.1038/479158a, http://www.nature.com/news/2011/111108/full/479158a.html, accessed 23 February 2014.
[29] ITU Radiocommunication assembly defers decision to eliminate the leap second, ITU Press Release, 19 January 2012, http://www.itu.int/net/pressoffice/press_releases/2012/03.aspx#.Uwpd-JWPPmQ, accessed 23 February 2014.
[30] Photo finish, Wikipedia, 7 December 2013, http://en.wikipedia.org/wiki/Photo_finish, accessed 24 February 2014.
[31] 100m - for the expert, 26 June 2002, http://www.iaaf.org/news/news/100-m-for-the-expert, accessed 24 February 2014.
[32] Fordyce, T., Usain Bolt Wins Olympics 100m final at London 2012, 5 August 2012, http://www.bbc.co.uk/sport/0/olympics/18907995,

Measuring the World, by John Austin

accessed 24 February 2014.
[33] Athletics abbreviations, Wikipedia, 18 ~December 2013, http://en.wikipedia.org/wiki/Athletics_abbreviations, accessed 24 February 2014.
[34] Reaction time test, Human Benchmark, http://www.humanbenchmark.com/tests/reactiontime/index.php, accessed 24 February 2014.
[35] Leap Second, the adjustment between atomic and Earth time, GreenwichMeanTime.co.uk, 9 April 2014, http://wwp.greenwichmeantime.co.uk/info/leap-second.htm, accessed 18 April 2014.

Measuring the World, by John Austin

Measuring the World, by John Austin

4. Frequency

Frequency = the number of cycles or events per unit time

4.1 Introduction

Frequency is a quantity derived from time. Essentially it is the number of events which occur in each unit of time. For example, the second is defined as a frequency (Chapter 3) rather than a duration. Thus time and frequency are intimately related.

In our every day lives, we are exposed to frequencies in music and sound, but also in our electrical power system and whenever we use electromagnetic radiation (e.g. Radio and television). These topics are mainly discussed in later chapters, since they are complex issues which require the prior development of other units, such as for energy and distance. Another area where frequency plays a role is in the disintegration of atoms, known as radioactivity. Radioactivity has possible health implications, and can also be used as a clock, for example to determine the age of the Earth. Computer chip speeds are also quoted as a frequency, with each calculation (multiplication, addition, subtraction, division) requiring several cycles.

In § 4.2, the units of frequency are discussed and used in § 4.3 and 4.4 to describe radioactivity and its health effects. The positive benefits of radiactivity are discussed in § 4.5 to 4.8 to date archaeological samples, and in § 4.9 to date the Earth itself. Other long period dating methods are described in § 4.10.

Measuring the World, by John Austin

4.2 Units of Frequency

As seen in Chapter 3, major unit systems use the second, and the corresponding frequency is sufficiently important to have its own name, hertz (abbreviation Hz). 1 Hz equals 1 cycle per second. Other units in use are revolutions per minute (an alternative phrase for cycles per minute) or cycles or revolutions over longer time periods, reserved for slow-acting processes. Although essentially the same sort of idea, radioactivity tends to be measures as disintegrations per second, which is often written s^{-1} rather than Hz.

Electricity in the home is delivered at 50-60 Hz, as discussed in Chapter 12. Electricity is in the form of Alternating Current, and the frequency quoted implies an oscillation from positive to negative and back to positive again 50-60 times per second. The frequency emerges from the method of generation. Higher frequencies occur in sound waves that we are receptive to, and middle C is at about 262 Hz[1]. Microwave ovens generate energy at typically 2450 MHz[2]. They are not actually tuned to the water vapour molecule, as is often mistakenly thought, but they work by forcing the vibration of any molecule which has a polar component[3]. Television and radio are broadcast at frequencies of about 100 MHz. Current computer chip speeds for personal computers are currently at about 3 GHz[4] and the number of transistors on integrated circuits have been increasing by about a factor of 2 every 2 years, known as Moore's law[5].

4.3 Radioactivity

The nuclei of some atoms are unstable, and disintegrate spontaneously with the emission of alpha (α), beta (β) or gamma (γ) radiation[6]. The fragments from the disintegration could themselves be unstable and decay further. α particles are

Measuring the World, by John Austin

helium nuclei, consisting of 2 protons and 2 neutrons combined. This radiation is usually easily absorbed, even by air molecules. β radiation consists of electrons, which are much more penetrating, while γ radiation is high energy electromagnetic radiation and most penetrating of all. We can never determine which specific atoms will disintegrate since the process is random. However, we can still get an accurate indication of the disintegration rate of a given sample, which is proportional to the number of atoms present. Thus a sample of given size will decay twice as fast (measured in terms of atoms decaying) as a sample half its size. This sort of process is common in nature and is referred to as exponential decay, as explained in Chapter 1. In mathematics,

$$N = N_o e^{-kt}$$

N is the number of radioactive atoms at time t, and N_o is the initial number. k is a constant specific to the radioactive element, and e is the base of natural logarithms $e = 2.71828.....$

Radioactivity was discovered by Marie and Pierre Curie, Henri Becquerel and William Roentgen at the end of the 19th century. Several of these scientists have units named after them. In 1896 Becquerel, the French physicist, found that rays from uranium ore fogged a photographic plate, similar to X-rays. 1 becquerel (Bq) is equal to 1 disintegration per second. This is a rather small unit, and measurements are often made in GBq (10^9 disintegrations per second) or TBq (10^{12} disintegrations per second). The Curies were the first to purify radium and observe its radioactive decomposition to radon gas. 1 curie is equal to 37×10^9 disintegrations s^{-1}, approximately the radioactivity of 1 g of radium. The curie is not a unit recommended by the SI because of the awkward factor 37. In 1903, Becquerel and the Curies received the Nobel prize for physics for their work on radioactivity.

The constant k, in the radioactive decay equation is usually expressed as a 'half-life' or the time taken for a given sample to disintegrate by a factor of 2. From the above equation, if

$N/N_o = 2$, then for the half-life $2 = e^{kt}$ or $t = log_e2/k \approx 0.6931/k$. Half lives range from fractions of a second to billions of years. If a sample has a half-life of 1 year, then after 1 year, half the radioactive element remains. After 2 years, one quarter remains, and after 3 years one eighth remains. The sample therefore gradually falls but never quite disappears entirely. Eventually, a handful of radioactive atoms remain, and those disintegrate randomly.

4.4 Radioactivity and Human Health

Different types of radiation have different health consequences on exposure. For human health, the SI unit is the gray (Gy), the radiation dose equivalent to an absorption of 1 J of energy per kg of exposed body mass. The unit is named after Louis H. Gray, the British biologist. The rad is the exposure when 1 g of body mass absorbs 100 ergs, and is a common unit in the cgs system. 1 gray = 100 rads. However, different radiation exposures produce different symptoms, reflected in the radiation 'quality factor'. The quality factor for β and γ radiation is set to 1. The quality factor for α radiation is about 10, and for neutrons it is 2-11, depending on their energy. By summing dose x quality factor, the equivalent dose is determined in so far as it is relevant to human health. This is measured in rems (Roentgen Equivalent in Man), when the dose is in rads, and sievert (Sv) when the dose is in grays. Grays and sieverts are part of the SI system. A typical chest X-ray delivers 0.08 mSv, a mammogram about 0.13 mSv and a whole body CT scan about 12 mSv[7].

The effects of radiation on human health are very complicated and barely a summary can be given here. The interested reader can start from a reputable source[8] and explore further. Doses > 1 Sv damage red and white blood cells. If the dose exceeds 3 Sv, death may follow within a few weeks. For doses > 10

Sv, cell linings and the digestive track die, and bacteria from the intestines invade the blood stream. Death occurs within about 1 week. With severe exposure, tens of Sv, brain injury occurs and death can come within hours. The treatment for mild exposures includes blood transfusions and the use of antibiotics. In Chernobyl in Ukraine, following the nuclear accident of 26 April 1986, hundreds of workers suffered radiation sickness and 28 died[7]. Typical doses from natural sources are about 1 – 10 mSv per year and about 240,000 people received a total dose exceeding 100 mSv over 20 years.

4.5 The Dating of Archaeological Samples with C-14

Radioactivity has transformed the dating of artefacts, rocks and minerals. Prior to the mid-20th century, dates for many objects were often poorly understood, based primarily on historical provenance: analytical techniques were not available. However, from 1949 onwards, Libby and co-workers[9] established a robust method of dating once-living objects, based on carbon-14 (C-14) content. The method has now proved to be an extremely valuable method of dating archaeological samples[10] and Libby received the 1960 Nobel prize for chemistry for his work.

Carbon-14 is an isotope of carbon with 6 protons and 8 neutrons (14 'nucleons') in the nucleus. It has a half-life of 5730 years, whereas carbon-12 (C-12) is stable, i.e. it is not radioactive. C-14 atoms decay by emitting an electron (the ionising radiation) and a neutrino, and one of the neutrons in the nucleus turns into a proton. This means that the product of decay is nitrogen-14 (N-14), or strictly a positive charged N-14 atom, N^+-14. The main point here is that the regular decay process can be used as a clock to age specimens. In particular, the method can be used to date previously living objects since all living objects contain carbon.

C-14 is produced in the upper atmosphere at a rate

which is approximately constant over decades. It mixes with atmospheric C-12 and is then ingested by plants. So, living plants tend to have a near constant C-14 to C-12 ratio. Plants are ingested by animals which soon acquire the same C-14 to C-12 ratio. However, as soon as the plant or animal dies, it no longer ingests fresh carbon. The ratio of C-14 to C-12 then decays away as the C-14 disintegrates. Therefore, if we find a piece of archaeological wood, we can compare the C-14/C-12 ratio with the value for living trees and calculate the age of the specimen. For example, the current C-14/C-12 ratio is about 1.0×10^{-12}[11]. Suppose that we measure the sample, and find the ratio to be 3.0×10^{-13}. If the age is t (years) then $3 \times 10^{-13}/1 \times 10^{-12} = e^{-kt}$, or $kt = \log_e 3.33$. Since the half-life of C-14 is 5730 years, $\log_e 2/k = 5730$. Hence, t = 5730 x $\log_e 3.33/\log_e 2$ = 9945, approximately.

By the 1950s, this technique proved highly effective for artefacts 3000-4000 years old. However, there was a tendency for 'radiocarbon' dates, as they are referred to, to be up to several hundred years younger than the most rigorously documented ages from the knowledge of the objects' history. The dating procedure assumed that natural C-14 was constant in concentration, but this was found to be incorrect. Over decades and centuries, C-14 does actually vary because it is produced by cosmic rays which vary substantially, depending on the activity of our own sun[12]. During high solar activity, cosmic rays are low, because the solar wind protects the Earth from particles from outside the solar system, the source of cosmic rays. The sun undergoes a near regular cycle, peaking every 11 years or so[13]. Once cosmic rays enter the upper reaches of the atmosphere, collisions with atmospheric gases generate C-14. There is considerable mixing in the atmosphere, so that by the time the C-14 reaches the surface the variations in production rate are much less significant, especially on a decadal time scale. Nonetheless, long-term corrections need to be made to the radiocarbon ages.

Measuring the World, by John Austin

These corrections are possible using data from tree rings. Essentially, in each year, trees create a new growth ring which is unique to the rain and temperature pattern of that year. By painstaking research, each individual tree ring can be counted and its C-14 content compared with that predicted for that year. In this way a graph can be plotted showing true age against radiocarbon age. Furthermore, by overlapping many tree records, it has now become possible to provide an accurate C-14 correction curve going back 12 ka (thousand years; a is the metric unit for year). See figure, taken from Reimer et al.[14]. One of the properties of the correction curve is that a given C-14 amount could correspond to either of two correct years. For example, a radiocarbon age of 3400 a could correspond to an actual date 3670 a or 3620 a, and further information would be needed to decide which year is more accurate. For the older age, 3670 a, the C-14 was relatively low because of a reduction in atmospheric C-14 production.

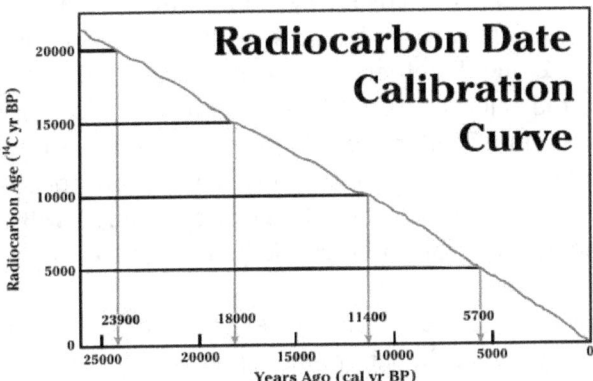

Comparison of Radiocarbon Dates to Calendar Dates[14]. Radiocarbon dates underestimate the actual age of the objects being dated, because the ratio of carbon-14 to carbon-12 has not been constant over time.

Measuring the World, by John Austin

The ambiguity of dates implies an inherent 1½ % systematic error in the date[14]. However, by combining several samples of slightly different ages, the errors can be reduced. One such example was given by W.L. Friedrich et al.[15], in which different depths of the same olive branch were used to date the eruption of Mt. Santorini between 1600 and 1627 B.C., implying about one third of the uncertainty of the date of a single sample.

Although levels of C-14 in the atmosphere (and hence ingested by living things), are primarily determined by cosmic rays, nuclear bomb testing in the 1960s produced C-14 which completely dominated the natural signal. Once testing ceased, C-14 has steadily decreased and is now near background levels. For a given C-14 amount, 2 dates are possible and additional information is needed. For example, by measuring C-14 in the tooth enamel of wisdom teeth and in the other teeth, 2 different C-14 levels are obtained, which can be used by forensic science to date a body to an accuracy of about 1 year or so[16]. This uses C-14 as a tracer, rather than using radioactive decay. For example, if in the wisdom teeth, the C-14/C-12 ratio is equal to that in the atmosphere in 1963 or 1972, and the other teeth in 1962 or 1975, then the earlier dates are the relevant ones, since wisdom teeth form when the person is about 17, after the other teeth have come through. So the hypothetical person above would have been born in about 1946. In 50 years, radioactive decay of an isotope with half-life 5730 years is too small to be considered, or at least only a small correction is needed.

One of the best known examples of C-14 dating is the "Turin Shroud", once thought to be the burial cloth of Christ. There is a negative image of a man on the cloth, as if the man had emitted a bright light after death. It is easy to see how religious enthusiasts would think of this as a sacred artefact. However, there is no record of its existence prior to about 1350, and this immediately raises suspicions as to its validity. Over the decades, the amount of material needed to date specimens decreased by orders of magnitude, and the church could no longer resist pressures to have it properly dated. In the 1990s 3 labs were selected to do the analysis and all the dates emerged consistently in the range 1260-1320[17], suggesting that it is indeed a medieval forgery. This particular artefact has been the subject of a tremendous amount of discussion, as can readily be found by entering 'Turin Shroud' in an internet search engine, and further discussion can be found in [16].

4.6 Dating with Radioactive Uranium

Uranium is a naturally occurring radioactive element. It provides much of the heat in the Earth's crust. The three isotopes U-238, U-235 and U-234 have vastly different half-lives of 4.468 Ga, 704 Ma and 245 ka, respectively[18]. U-238 is the most common form and can be used to date the Earth's rocks. The neutrons from U-235 decay can be used in a nuclear reactor in a self-sustaining reaction, whereas U-238 absorbs neutrons. However, U-235 needs to be increased as a fraction of total uranium from 0.72% background level to about 5%. Isotopes in natural uranium can be separated with a gas centrifuge[19]. It is usual to have a bank of centrifuges, as many as 10, each producing more refined U-235 in sequence. The method first produces uranium hexafluoride gas, UF_6. The method takes advantage of the different separation rates of different molecules, depending on their molecular mass. Fluorine has an atomic weight of 19, so the 6 fluorine atoms together have mass 114. Hence, the different uranium isotopes produce UF_6 gas with molecular weights 114 + 238 = 352, and 114 + 235 = 349. The separation rates vary as the molecular weight, so each stage of the centrifuge can enhance the U-235 by a factor 352/349, or about a factor 1.01. The machines are very large, about 10 m high, i.e. several stories high, and about 2 m in diameter. Fifteen of these machines, typically rotating at 600 m s^{-1} at their outer edges will produce an eventual enhancement of U-235 ratio to about 4%[20]. The equipment is highly technical and enabled by specialised software; the secrecy surrounding nuclear technology means that technical details are scarce. In June 2010, the USA is thought to have introduced a computer virus into the software which ran Iran's uranium enrichment centrifuges[21], which essentially put the equipment out of commission for over a year.

Igneous rocks can be aged using the U-238 decay curve, but a better approach is to measure the concentrations of specific lead isotopes. Igneous rocks are produced from the cooling of molten material. At these high temperatures, either the lead will have been boiled off, or it would have sunk below the rocks themselves. So at the moment of crystallisation, igneous rocks are free of lead. Uranium decays in several stages, finally reaching lead as a stable element. Therefore, the presence of lead in rock arises from the decay of its uranium. U-238 decays to lead-206 (Pb-206) and U-235 decays to Pb-207. The first step in both cases is much slower than the subsequent ones, so we can write

$$Pb_{206} = U_{238}(1 - e^{-k_1 t}) \quad \& \quad Pb_{207} = U_{235}(1 - e^{-k_2 t})$$

and the equations for the uranium isotopes are

$$U^t_{238} = U_{238}\, e^{-k_1 t} \quad \& \quad U^t_{235} = U_{235}\, e^{-k_2 t}$$

These equations represent the slow exponential growth of lead as the uranium decays. From this, and using the current ratio of U-235 to U-238, U^t_{235}/U^t_{238}, of 0.0072 we obtain

$$\frac{Pb_{207}}{Pb_{206}} = \frac{0.0072\,(e^{k_2 t} - 1)}{(e^{k_1 t} - 1)}$$

Using the half-lives of 0.704 and 4.47 Ga for U-235 and U-238 respectively, to substitute for k_1 and k_2, we can compute t, if we can measure the ratio of the lead isotopes on the left. This can be determined by placing the sample, or a small part of it, in a mass spectrometer[22].

4.7 Potassium-Argon Dating

Potassium-40 (K-40) can be used to date sedimentary rocks. Sedimentary rocks are laid down a thin layer at a time and the layers harden from the weight above. Potassium-40 disintegrates into argon-40 (10.72%) or calcium-40 (89.28%)[23]. The argon is released into the atmosphere at the top layer. In deeper layers, the argon is trapped in the rock. As K-40 is the only source of A-40, by measuring the Ar-40/K-40 ratio, the rock can be dated. This ratio is given by $e^{kt} - 1$ where t is the age of the rock, and k is the constant of radioactive decay, equivalent to a half-life of 1.248 *Ga*. To reduce analysis uncertainties, the sample is irradiated in a nuclear reactor. This converts the potassium to Ar-39. The Ar-40/K-40 ratio in the original sample is the same as the Ar-39/Ar-40 in the irradiated sample, and the ratio can be determined more accurately in a mass spectrometer[16]. The technique can be used to date rocks, which gives the dates of fossils embedded in the rocks. Although the precision of measurements has improved substantially over time, there is always a need for higher precision so that key evolutionary questions can be addressed. For example, how quickly does natural selection work to create new species, or make others extinct? The date of the disappearance of the dinosaurs at the K-T boundary[24], 65.5 Ma ago, is now known to a precision of 0.1 Ma, but that is still not high enough precision to determine whether they became extinct suddenly (as hypothesised) or gradually by natural evolution. On the plus side, K-Ar dating has demonstrated that the solar system formed over a relatively short period (< 10 Ma), compared with a previously hypothesised 100 Ma.

4.8 Summary of Nuclides Used for Dating and Testing

In the following table is presented a brief summary of radioactive nuclides that are useful for dating a range of specimens. Each particular technique has its own area of specialisation depending on the half-life of the nuclide used. Some of these methods have already been discussed. In addition the tree ring measurement is indicated for calibration of the other dates.

Nuclides used in age determinations

Technique	Age range	Precision (approx.)	Notes
Tree rings	0-12 ka	1 a	Standard for radio-carbon adjustment
Radiocarbon	50-50 ka	30 a (age < 5 ka)	Correction with tree ring data
C-14 from bombs	0-50 a	1 a	C-14 used as a tracer, not radioactive decay curve
Str-87/Str-86	100 ka – 500 Ma	50 ka	Estimates based on [25]
K-Ar or Ar-Ar	50 ka – 2 Ga	0.1 Ma	Analysis of sedimentary rocks
U-Pb or Pb-207/Pb-206	100 Ma – Earth age	0.6 Ma	Analysis of igneous rocks

Amongst other radioactive disintegrations useful to archaeology is the ratio Strontium-87 to Strontium-86 (Str-87/Str-86)[25]. Like the chemically similar element calcium, strontium is incorporated into plants and animals. Strontium comes from the decay of rubidium and its concentration could be used both to date samples, and to identify their source region. This was used for example to identify the source region in the case of the Chaco ruins in New Mexico and has been used to date nests (middens) of pack rats[26]. Diamond[26] also identifies other archaeological uses for radionuclides. Isotopic composition of water – the O-18/O-16 ratio is enhanced in animals during drought conditions and this has been used to conclude that environmental degradation (deforestation) led to the collapse of the Maya civilisation. Radiocarbon analysis has also been used to determine the fraction of sea food to land food in the diet of the Greenland Norse to provide insights into their disappearance in the late 1400s[27].

4.9 The Age of the Earth

Prior to the discovery of radioactivity, the age of the Earth could only be estimated, and the values obtained were much lower than current values. In 1862, Lord Kelvin calculated a value of 20 – 400 Ma[28], with a most likely value near 100 Ma. The age was determined by calculating the heat from a deep mine to the surface. If radioactivity and the associated heat production had been recognised a much longer age would have been accepted. Despite this discrepancy, over a period of decades, scientific discussion extended the most likely age to the lower limit, rather than the higher limit. This age contradicted Darwin's recently proposed theory of natural evolution, as 20 Ma would likely have been insufficient time for natural selection to have worked. By 1905, the appreciation of radioactivity extended the most likely age to as high as 500 Ma, still well short of current thinking. In 1907, the decay of uranium to its final stable element, lead, was used to place the age of the Earth in the range 400 – 2200 Ma[29]. However, the half-life of uranium was not well known. By the 1950s the ratio of the two stable lead isotopes described above eventually gave an age of 4.55 Ga[30], close to the currently accepted value. This required substantial and careful analysis as by this time, lead had become ubiquitous in the air and on exposed surfaces. The high concentrations of atmospheric lead arose from its use as an anti-knocking agent (tetra-ethyl lead) in petrol engines. Lead was also used extensively in paints. It's a bizarre aspect of fate that Midgley was the inventor of both tetra-ethyl lead, which poisoned the environment, and chlorofluorocarbons (CFCs), which nearly destroyed the ozone layer[31]. Although in the developed world, environmental lead has been substantially reduced, the residues of the CFCs will remain in the atmosphere for some decades more, and continue to cause the annual spring Antarctic ozone hole!

Using iron meteorites to date the solar system, analyses have settled into the age range 4.51 – 4.56 Ga, for the age of the Earth, as the half-life of uranium-238 became better known. It is reasonable to assume that the ages of meteorites match that of the Earth. Analysis of moon rocks gave a similar figure[32] suggesting a common origin. Current values provide an even more precise figure for the age of the solar system, 4.5682 ± 0.0006 Ga[33].

4.10 Other Timing Methods

Periodically, the Earth's magnetic field reverses polarity, imprinting its magnetism on rocks. The reversals take a few ka to complete and occur on average about once every 450 ka[34]. The reversals appear quite random, but can be calibrated with K-Ar techniques. Hence, samples can be measure for their K-Ar and magnetisation and then if a second sample is discovered with known magnetisation patterns, the second sample can be dated by reference to the earlier K-Ar analysis.

Other cycles have also been imprinted into rocks and ice core data. For example, Milankovitch cycles are changes in the Earth's orbit over periods of 21 ka[35]. By examining ice core data

Measuring the World, by John Austin

climate changes have been determined[36]. Essentially, changes in total solar irradiance and atmospheric constituents such as CO_2 have changed the climate. These observations have encouraged climate change denialists to claim that climate change is nothing to be concerned about since it occurred in the past. However, this is to ignore entirely past changes in the Earth's orbit around the sun, whereas current changes are due to an entirely different process – the accumulation of human produced greenhouse gases.

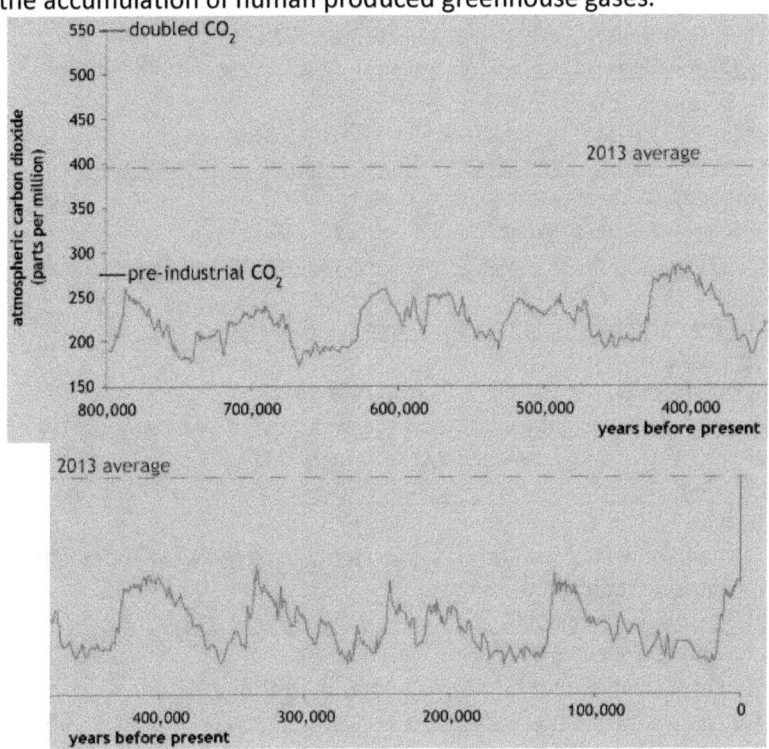

The rise of CO_2 since the industrial era is much larger than previous changes, and it has occurred very suddenly on a geological time scale. In the diagram above, the CO_2 amount is shown as a function of time, measured from ice core data[37]. The record goes back almost a million years, during which the CO_2 amount oscillated between about 175 and 275 parts per million by volume (ppmv), but since 1950, the concentration has risen to almost 400 ppmv, which appears to be sudden on a geological time scale.

References

[1] C (musical note), Wikipedia, 17 February 2014, http://en.wikipedia.org/wiki/C_(musical_note), accessed 24 February 2014.
[2] Nave, R., Hyperphysics, Electricity and Magnetism, http://hyperphysics.phy-astr.gsu.edu/hbase/waves/mwoven.html, accessed 24 February 2014.
[3] Microwave oven, Wikipedia, 24 February 2014, http://en.wikipedia.org/wiki/Microwave_oven, accessed 24 February 2014.
[4] CPU Benchmarks, 2014, http://www.cpubenchmark.net/, accessed 24 February 2014.
[5] Moore's Law, Wikipedia, 23 February 2014, http://en.wikipedia.org/wiki/Moore's_law, accessed 24 February 2014.
[6] Radioactive decay, Wikipedia, 24 February 2014, http://en.wikipedia.org/wiki/Nuclear_radiation, accessed 24 February 2014.
[7] Health effects of the Chernobyl accident: an overview, World Health Organisation, April 2006, http://www.who.int/ionizing_radiation/chernobyl/backgrounder/en/, accessed 25 February 2014.
[8] Acute radiation syndrome, Wikipedia, 14 February 2014, http://en.wikipedia.org/wiki/Acute_radiation_syndrome, accessed 25 February 2014.
[9] Arnold, J. R. and Libby, W. F. ,1949: "Age Determinations by Radiocarbon Content: Checks with Samples of Known Age," *Science,* **110** (2869), 678–680.
[10] Radiocarbon dating, Wikipedia, 16 February 2014, http://en.wikipedia.org/wiki/Radiocarbon_dating, accessed 25 February

Measuring the World, by John Austin

2014.
[11] Carbon-14, Wikipedia, 23 January 2014,
http://en.wikipedia.org/wiki/Carbon-14, accessed 25 February 2014.
[12] Cosmic Ray, Wikipedia, 22 February 2014,
http://en.wikipedia.org/wiki/Cosmic_ray, accessed 25 February 2014.
[13] Solar variation, Wikipedia, 18 February 2014,
http://en.wikipedia.org/wiki/Solar_variation, accessed 25 February 2014.
[14] Reimer, P.J. et al., 2004: IntCal04 Terrestrial radiocarbon age calibration, 0-26 cal kyr BP, Radiocarbon, 46, 1029-1058.
[15] Friedrich, W.L. et al., 2006: Santorini eruption radiocarbon dated to 1627-1600 B.C., Science, 312 (5773), 548.
[16] D. McDougall, 2008: "Nature's clocks: how scientists measure the age of almost anything", University of California Press, 2008.
[17] Damon, P.E. et al., 1989: Radiocarbon dating of the Turin shroud, Nature, 337, 611-615.
[18] Isotopes of Uranium, Wikipedia, 17 February 2014,
http://en.wikipedia.org/wiki/Isotopes_of_uranium, accessed 25 February 2014.
[19] Gas centrifuge, Wikipedia, 6 January 2014,
http://en.wikipedia.org/wiki/Gas_centrifuge, accessed 25 February 2014.
[20] Kessler, G., Proliferation-proof uranium/plutonium fuel cycles : Safeguards and non proliferation, KIT Scientific Publishing, 2011.
[21] Stuxnet, Wikipedia, 25 February 2014,
http://en.wikipedia.org/wiki/Stuxnet, accessed 25 February 2014.
[22] Hoffman, Edmond de & Stroobant, V., Mass Spectrometry: Principles and applications, John Wiley and Sons, 2013.
[23] Potassium-40, Wikipedia, 7 February 2014,
http://en.wikipedia.org/wiki/Potassium-40, accessed 26 February 2014.
[24] Dinosaurs, Smithsonian National Museum of Natural history,
http://paleobiology.si.edu/dinosaurs/info/everything/why.html, accessed 26 February 2014.
[25] McArthur, J.M., Howarth, R.J. & Bailey, T.R., 2001: Strontium isotope stratigraphy: Lowess version 3. Best fit to the marine Sr-isotope curve for 0 to 509 Ma and accompanying lookup table for deriving numerical age, Journal of Geology, 109, 155-170.
[26] Diamond, Jared, "Collapse: how societies choose to fail or succeed", published by Viking Penguin Group, 2005.

[27] Arneborg, J. et al., 2002: C-14 dating and the disappearance of Norsemen from Greenland, Europhysics News May/June 2002, 33 (3), 77-80.
[28] Age of the Earth, Wikipedia, 28 January 2014, http://en.wikipedia.org/wiki/Age_of_the_Earth, accessed 26 February 2014.
[29] Boltwood, B. B.,1907: On the ultimate disintegration products of the radio-active elements. Part II. The disintegration products of uranium, *American Journal of Science,* **23** (134), 77.
[30] Patterson, Claire, 1956: Age of meteorites and the earth, *Geochimica et Cosmochimica Acta,* **10** (4), 230–237.
[31] Thomas Midgley, Jr., Wikipedia, 23 February 2014, http://en.wikipedia.org/wiki/Thomas_Midgley,_Jr., accessed 26 February 2014.
[32] Moon rock, Wikipedia, 24 February 2014, http://en.wikipedia.org/wiki/Moon_rock, accessed 26 February 2014.
[33] Bouvier, A. & Wadhwa, M., 2010: The age of the solar system redefined by the oldest Pb-Pb age of a meteoritic inclusion, Nature Geoscience, 3, 637-641, Published online 22 August 2010, doi:10.1038/ngeo941.
[34] Geomagnetic reversal, Wikipedia, 14 February 2014, http://en.wikipedia.org/wiki/Geomagnetic_reversal, accessed 26 February 2014.
[35] Milankovitch cycles, Wikipedia, 24 February 2014, http://en.wikipedia.org/wiki/Milankovitch_cycles, accessed 26 February 2014.
[36] Ice core, Wikipedia, 23 February 2014, http://en.wikipedia.org/wiki/Ice_core, accessed 26 February 2014.
[37] Lindsey, Rebecca, 2014: How much will Earth warm if carbon dioxide doubles pre-industrial levels, NOAA Climate.gov, 24 January 2014, http://www.climate.gov/news-features/climate-qa/how-much-will-earth-warm-if-carbon-dioxide-doubles-pre-industrial-levels, accessed 26 February 2014.

5. Length

Length = the shortest distance between two points

5.1 Introduction

Determining length is one of the most important measurements in society as well as in science. Length units therefore have a very long history (§ 5.2-5.3). Most of that history concentrates on the scales that are mainly observable by humans. Outside that range, powerful instruments have been developed to explore the atomic scale and the astronomical scale, discussed in Chapter 15.

Length is the shortest distance between two points, which may seem obvious. We generally assume 'Euclidean Geometry' in which the shortest distance is a straight line. However, the general relativity (GR) theory of Einstein shows that massive objects curve space itself, so that the shortest distance follows a curve. The curvature is usually minuscule, but it is sometimes detectable.

Modern communication is crucially dependent on the details of the electromagnetic spectrum, often described by wavelength as explained in § 5.6. Another important aspect of length or distance is navigation. Obtaining a latitude fix on the Earth's surface has always been relatively straightforward by observing the position of astronomical bodies (sun, moon, stars etc.). However, it is not possible to obtain longitude in this way. The problem was finally solved by the invention of high quality clocks (§

5.7). The modern method is to use satellites and the global positioning system (GPS) to obtain a position by triangulation. Accurate positioning depends on GR theory, so the concept of curved space is not just a theoretical construct, although the degree of curvature induced by the Earth is extremely small. As well as latitude and longitude, GPS can also be used to determine accurate altitudes for mountains, or even the exact topography of the Earth. This replaces triangulation methods that used to be extremely laborious to apply. Furthermore, GPS methods can detect changes in topography, in near real time, due to earthquakes and volcanic eruptions for example. Related to navigation is the scaling of maps, discussed in § 5.8.

Careful length measurement plays an important role in diagnosing vision impairment, discussed in § 5.9. Length and time are combined in § 5.10 and 5.11 which explore speed and velocity in general, and atmospheric wind in particular. Finally, in § 5.12, distance measurement in sport is considered. There are many other examples of distance measurement in society, information about which can be obtained from other sources.

5.2 The History of Units: the Human Body

Investigations suggest that ancient civilisations such as the Romans[1] and Egyptians[2] had a basic set of units for length These were often based on the human body. Starting with the cubit, which is the distance from the tip of the fingers to the elbow, we have evidence that the Israelite cubit, at the time of Solomon in around 2000 BC was 640 mm[3]. For the Ancient Egyptians, a royal cubit was 523 mm and for the Ancient Romans 444 mm. My measurement is 452 mm, so I would have been about the same size as a Roman and not big enough to have been an Egyptian Royal or an Israelite!

The foot also began in ancient times, probably based

Measuring the World, by John Austin

on the length of the human foot. By comparing size of ancient measuring rods, with modern measurements we can deduce that 1 Roman foot = 0.296 m, very similar to the current definition of the fps foot, 0.3048 m. By the middle ages, in different European countries, the foot varied between 25 and 50 cm (10 and 20 inches). In 1305, in the reign of Edward II, England set the foot at 12 inches (0.3048 m) where 1 inch equalled "the length of 3 grains of barley, dry and round"[4]. This length standard still remains in the fps system. The Romans also had a measure for 1 mile which was 1000 "paces" where 1 pace is a double stride or 5 Roman feet. The origin of the yard is a bit uncertain and may have been set at 2 cubits.

When I was in my early teens, I became fascinated with the use of the body for providing a set of length scales. You can obtain a surprisingly high precision estimate if you have already calibrated your body against true measurements. An inch for example is approximately the length of the last segment of the thumb, excluding the nail. Likewise, you can assume the above ancient measurements for cubit (45 cm), foot (30 cm) and stride (90 cm). The last unit is not very robust as it depends on how much you are prepared to stretch. Most people's natural walking stride length is about 75 cm (2 ft 6 in). I still use my hands to estimate lengths when I don't have a measure handy (pun intended). The width of the hand is about 12 cm (5 in, including the width of the thumb) and its length about 20 cm (8 in). It seems I'm not the only eccentric one[5]. Another important measure is the distance between your fingertips with outstretched arms, which is close to your height, as noted by Leonardo da Vinci in about 1490[6]. In fact this turns out not to be as reliable as I once thought. For example, for the swimmer Michael Phelps, the distance is 2.03 m compared with his height of 1.93 m. This small (5%) difference (as well as big feet!) perhaps has contributed to him being a world-class swimmer[7]. On one occasion in my early teens, as part of a scouting trip, I was asked to hike to a particular bridge and to mea-

Measuring the World, by John Austin

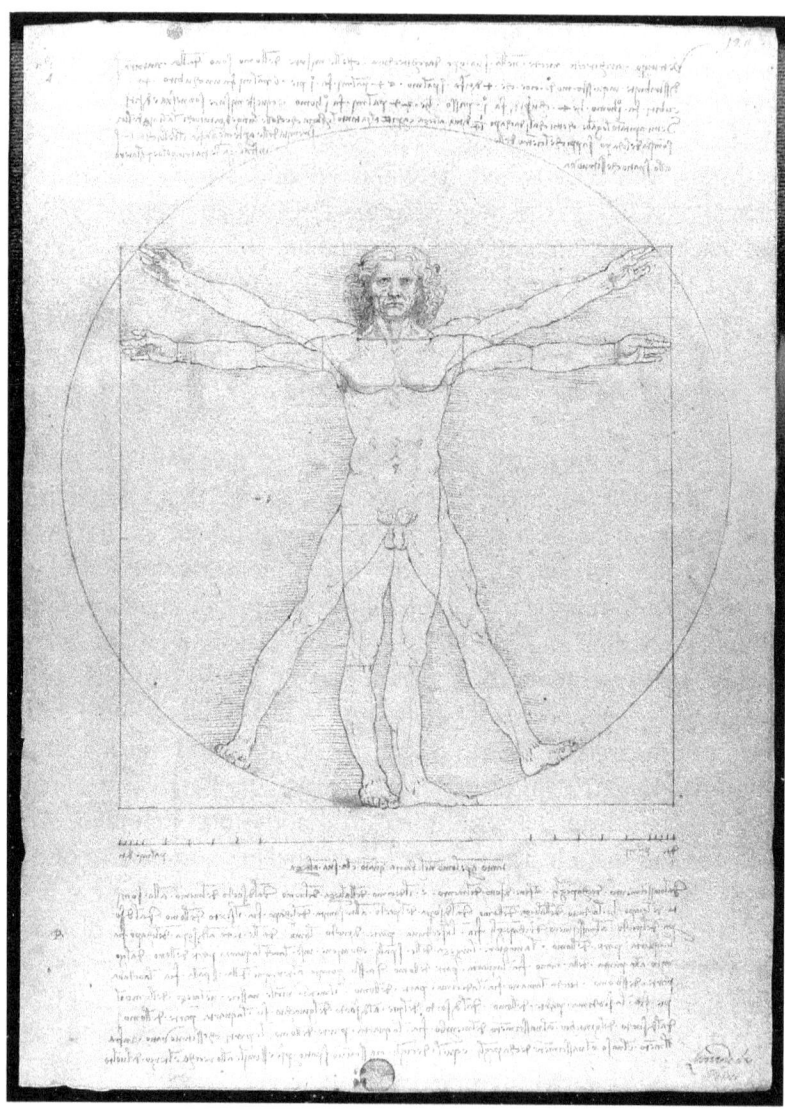

The Vitruvian man of Leonardo da Vinci sketched in about 1490. See the article in Wikipedia [6].

sure it, without benefit of a tape measure. I used my hands and came up with a measurement of 10 ft 6 in. Others came up with larger distances because they had used their uncalibrated stride length, whereas my interest in the human body ensured that I had already calibrated my hand size. Of course since that time I have grown into an adult, my dimensions have changed (with the hand length going from 15 to 20 cm), and I have abandoned fps units. Each person is of course a slightly different shape and so the above general units are not robust. Nonetheless, for a single individual prior calibration of the body against a tape measure can allow measurements to be made which are accurate to within a few %. One can speculate that the awkwardness of fps units (12 in per foot, 3 feet per yard) comes from the shapes of our bodies rather than any underlying plan. Although the metric system may not have as many *ad hoc* units available, it scores in many other ways, not least the fact that carrying out calculations is nearly always considerably easier.

A league is another old unit which appears in the classic literature. It was traditionally a distance of about 3 nautical miles or about 5.5 km[8], but varied quite a bit. In the Jules Verne classic, a league was about 4 km[9], so 20,000 leagues is indeed a bit deep and in fact about 12 times the Earth's radius!

5.3 The history of Modern Units

The primary unit in the fps system is the foot, with 12 inches per foot. A number of exotic units relate to the foot or yard. Many of these relate to farming before the industrial revolution. Despite the absence of old farming methods, some of these units persist, including the chain (22 yd) and furlong (220 yd). See Chapter 6. Some fancy arithmetic is sometimes needed to convert from one unit to the next, but then that's a consistent theme with the fps units.

Measuring the World, by John Austin

The base length unit in the metric system is the metre, which was originally defined as one ten-millionth of the distance from the pole to the equator of the meridian arc passing through Dunkerque and Barcelona[10], assumed to be equal to the meridian arc through Paris. This would imply a polar circumference of 40,000 km. Modern measurements place the Earth's polar circumference at 40,007.863 km[11], compared with the equatorial circumference of 40,075.017 km. The diameters pole-to-pole and across the equator are 12,713.505 and 12,756.274 km respectively. This gives a difference of 42.769 km, or about 1/298 of the diameter. Thus the Earth bulges slightly at the equator. In fact the circumference is actually greatest just south of the Equator, and the Earth is slightly pear-shaped. Moreover, the exact measurements vary in time due to the weight of ice pressing down on the Earth's crust ("post-glacial rebound") as well as other reasons including earthquakes, plate motion and atmospheric pressure, but the main factor is climate change[12]. Although the Earth does not provide a stable standard, the meridian size could not at the end of the 18th century be measured to better than 1 part in 10^4[13]. Later, the yard and the metre were retained as the distance between points marked on metal bars kept in laboratories in clean room conditions. This enabled the metre to be reproduced to an improved precision of 1 part in 10^5 by the end of the 18th century, and 1 part in 10^7 by the end of the 19th century. The yard has a rich history[14] but physical standards have a problem, as was found in 1950 when it was reported that the imperial yard had been shrinking at a mean rate of about 0.4 parts per million per decade[15], as noted in § 2.1. The metre also has a rich, if shorter history[16] and the physical standard having been constructed from a platinum-iridium alloy was much less prone to drift, although there could be no certainty what that drift rate might be. Restrictions with the use of the physical artefact definition of the metre led in 1960 to its redefinition in terms of the wavelength of

Measuring the World, by John Austin

the emission of radiation from krypton-86[17].

At this stage, the independence of the foot was abandoned, and instead set to 0.3048 m exactly. The krypton standard allowed the precision of the metre to be set to 1.3 parts in 10^9[13: Note that this reference records the precision of 4×10^{-9}, but it is actually a 3 standard error value quoted from the reference 18], the precision at which the wavelengths could be defined and counted. By the 1980s, lasers had become established as a stable light source. Moreover special relativity theory of Einstein and Lorenz demonstrated that the velocity of light was the same to all observers. The velocity of light was already known to 1 part in 10^9, and hence a higher precision was limited by the precision that the metre was known. In 1983, the metre definition was redefined again[18] by specifying a fixed velocity of light in a vacuum: 299,792,458 m s^{-1}. Hence the metre is now the distance travelled by light in 1/299,792,458 seconds. This has enabled the precision of the metre to be improved from 1 part in 10^9 to 2.1 parts in 10^{11}[19], the precision of the reproducibility of the frequency standard used for distance measurement. Moreover, since radar beams (which travel at the speed of light) are used in GPS positioning, times of beams are exactly equivalent to distances.

The modern set of length units depend on advanced physics: the special theory of relativity (SR). SR was stimulated by unexplained results in the measurement of the velocity of light. The Michelson-Morley (MM) experiments measured its velocity in 1887 in what are now classics in physics experimentation[20]. Measuring the speed of light is technologically challenging because of its high value. In the MM experiments, a bright source of light is used to provide a narrow beam which is reflected around a rectangular-shaped region by a sequence of mirrors. The mirrors rotate at a known frequency and the light can only complete the circuit if all the mirrors are aligned at the instant the light beam strikes. This then puts tight constraints on the time taken for the light beam to traverse the rectangle. From the timings and the distances between the mirrors, the velocity of light can be determined. The MM result was a measured velocity of 299,796 km s^{-1} in modern units, an error of only 4 km s^{-1} (a fractional error of 1.3 x 10^{-5}, or 0.0013%). The puzzling aspects of the MM results at the time was the consistency of values obtained at different times of year. The Earth would have been at different points in its orbit and would have been travelling at different velocities relative to background space. This velocity at different times of year would have been ± 30 km s^{-1}, easily detectable by the MM experiments. This lack of variation of c, the velocity of light, remained unexplained until the Einstein-Lorenz theory of special relativity in 1905. This theory [21, for a modern exposition], is referred to as the 'special' theory of relativity since it relates to observers which are not accelerating relative to each other, but moving at constant velocity. The theory demonstrates that all observers are of equal status and that there is no special framework (background space: the aether), with which to compare observations. Further, there is a universal speed limit of c: no objects of finite mass can accelerate to a velocity equal to or greater than c. The theory predicts how mass and length change as the velocity of an object approaches c, and a number of exotic phenomena occur (time dilation, length contraction). Our senses do not naturally accept these conclusions, but the predictions of special relativity have been fully tested and found to be accurate. Rather, our intuition expects dynamics to follow the predictions of Newton first published in 1687, but such laws strictly follow only if the speed of light is infinite. For most purposes, the velocity of light is so large that it might as well be infinite, so Newtonian dynamics are quite accurate enough for the majority of situations. Newtonian dynamics are simply the limiting equations of special relativity, for infinite c.

Measuring the World, by John Austin

5.4 Conversion of Units

The conversion between metric and imperial units, and the converse, is given in Appendix B. For an approximate conversion from m to yd, add 10% and the error will only be 0.7%. To convert from yd to m, subtract 10%, for an error of 1.6%. For example, 100 yd is 91.44 m, about 90 m, while 100 m is 109.36 yd, or about 110 yd. One mile (8 furlongs) is 1609.344 m (exactly).

Ironically, Estate Agents often use laser measuring devices to size up individual rooms. By the definition of the metre, the timing beam gives the distance directly in metres, yet the distances are often quoted in feet!

5.5 Peculiar Units

As if the fps system were not peculiar enough, there are several other units in occasional use. One rod is defined as 16 ft 6 in, which sounds strange enough, until you see that an acre is sized 4 x 40 rods (Chapter 6). Water depth for no rhyme or reason that I can see is often measured in fathoms (6 ft), I just can't fathom it! A traditional method of measuring the height of a horse is using the width of a hand. This is a rough and ready method (§ 5.2), but a hand is now officially exactly 10.16 cm (4 in).

Other silly units used by particular professionals include the US Survey foot ($1200/3937$ m ≈ 0.3048006 m)[22]. Why does it always seem to be the engineers (especially US ones) who have questionable ideas? It's the civil engineers this time, and their foot is almost indistinguishable from the real foot (0.3048 m), so why do they feel the need to have their own? As an example, a distance of 10 million feet, 3000 km in conventional units, about the size of the USA, would be surveyed as just 6 metres different to conventional feet. If after all that, it really does matter to distinguish between conventional and US survey feet, wouldn't it be better simply to use metres which come at an agreed size? That would eliminate all confusion. A skein of yarn is 360 ft, which isn't much different to 100 m, so why not give up questionable habits? A 'line' in tailoring is 1/40" and typewriters or computers define point size as 0.3515 mm or 1/72" (0.3528 mm), depending on who you talk to. Use of these units puts up barriers to communication and one wonders whether this is the purpose of some professions. Of course all or some of the above units may be extinct by now and it may be that they exist only as curios in the dictionary or encyclopedia.

5.6 The Electromagnetic Spectrum

Light is a form of electromagnetic (EM) wave and as indicated in § 5.2 is now used to define the metre. However, EM waves are all-pervading in society and all travel at the same speed (in a vacuum this is often denoted c, although the velocity is slower in other media such as air depending on the refractive index). The waves arise from oscillating electric and magnetic signals (Chapter 12) at right angles, or transverse to the direction of travel of the waves. In that respect they are quite different from sound waves (Chapter 10) which arise from oscillations along the direction of travel. Newton demonstrated in the 1660s[24] that If you take white light and pass it through a prism, you get a range of colours from red to violet. A rainbow occurs in the sky when raindrops act as tiny prisms. The colours we see are the brain's response to triggering at different amplitudes of the 3 optical receptors, which are sensitive to red, green and blue light. Different birds and insects have different colour vision which determines the sort of flowers that are pollinated[25]. Bees can't see red, so they tend to pollinate yellow and blue flowers with a sweet odour. Many yellow flowers have UV markings, which of course we can't see but bees can. Butterflies also pollinate yellow or blue flowers. Beetles and

Measuring the World, by John Austin

Wave	Wavelength (m)	Frequency (Hz)	Common Use
Gamma rays	10^{-11}	3×10^{19}	Radiation therapy for cancer
X-rays	10^{-10}	3×10^{18}	Medical imaging
Ultra-Violet (UV)	$< 0.36 \times 10^{-6}$	$> 0.83 \times 10^{18}$	Sun-bathing, industrial antibiotic
Visible (light)	$0.36 - 0.72 \times 10^{-6}$	$0.42 - 0.83 \times 10^{18}$	Sight
Infra-red (IR)	$> 0.72 \times 10^{-6}$	$< 0.42 \times 10^{18}$	Cooking, heating
Microwave (MW)	10^{-3}	3×10^{11}	Cooking, GPS, mobile phones
Radio waves	10^{-1}	3×10^{9}	TV, Radio broadcasting

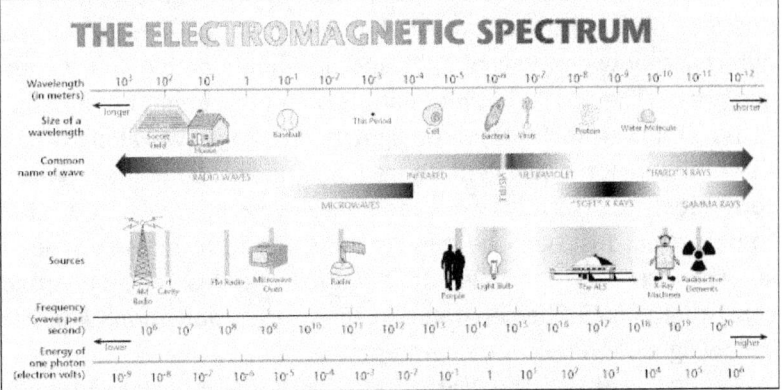

The electromagnetic spectrum, from Lawrence Berkeley Laboratory, USA[23].

flies pollinate white flowers with spicy odours, e.g., magnolia and wild roses. In tropical locations, humming birds pollinate preferentially red, orange, and yellow flowers – for example, columbines and fuchsias, but they have good colour vision across the full range and beyond ours due to their additional receptors[26].

We call colours red, green and blue *primary* as human vision can separate these wavelengths [27]. There is nothing fundamental about it. In other words, an alien race might define primary in a different way. Even a dog would, if it had the smarts! In principle, we could have evolved with more colour sensitivity (e.g. into the UV or better resolution in the red), or even less sensitivity. Perhaps we could have had a world without red at all. The fact that we are strictly sensitive to only three colours means that in principle we can be easily fooled. For example, if we saw yellow light, it could indeed be yellow, or it could alternately be a combination of equal amounts of red and green. We have no means of distinguishing between them since our vision would give an equal reaction in the blue and green colour receptors in both cases. In other words, it would only be by analysing the light with a prism or equivalent instrument that we would be able to tell the difference between yellow light and an equal combination of red and green. There is sometimes confusion on this point, but it needs to be emphasised that yellow light has a single wavelength and that this is not equivalent to red and green combined, which continues to have two separate wavelengths after the combination. Just because it looks the same to us doesn't make it so. The use of a prism would demonstrate this, since the two constituents could be separated again after combination, whereas yellow light cannot be so separated.

Outside the visible range, although not detectable by our own bodies, EM waves are still very important in society, covering the spectral range from gamma rays to radio waves. Chapter 15 describes how our observations of the full range of the EM spectrum has advanced our understanding of the universe.

Television is a special case of radio waves. The audio signal is carried by frequency modulation (changes in frequency of the carrier signal), and the video signal is carried by changes in

Measuring the World, by John Austin

wave amplitude (amplitude modulation, AM). The UHF signal in the range 300 MHz – 3 GHz can provide many tens of TV channels, and the precise frequency usage varies from one country to the next[28]. Signals are broken into the three primary colours blue, green and red. TV signals are converted to microwave wavelengths to send from stations to relay towers and between relay towers.

Radio broadcasting is similar. FM radio[29] uses a fixed amplitude oscillation, and a frequency variation relative to the carrier wave. AM radio[30] has fixed frequency and varying amplitude. FM is free of the static from thunderstorms, and more faithfully reproduces music and speech.

For domestic cooking, microwave ovens use wavelengths of about 10 cm. The ovens have metal reflecting walls and a rotating stirrer to ensure uniform amplitude waves throughout the cooking cavity. Most ovens have a rotating food plate also to enable food to cook more evenly. Most ovens have a viewing window, consisting of a metal plate with hole a few mm wide. The size of the hole is much less than the wavelength of the microwaves, so the waves are reflected by the remaining metal in the plate, but there are enough holes placed close enough together to be able to see clearly inside. Incidentally, this trick exploits our binocular vision. The magnetron, which produces the microwaves, typically has a frequency of 2.45 GHz for domestic ovens[31] and the microwaves are absorbed by molecules such as water which have a dipole charge (i.e. a charge separation at the two ends of the molecule). Virtually all foods contain some water. However, in ice, the water molecules are less free to move and don't absorb the microwaves as effectively. So ice itself and frozen foods do not thaw out readily in a microwave oven. Incidentally, this is exploited in a silly (but fun!) recipe for 'reversed baked Alaska' (also known as a frozen Florida for obvious reasons if you appreciate the climate in the USA), in which the ice cream remains frozen on the outside, while the cake is cooked on the inside[32]. Compared with a conventional (infra-red) oven, microwaves have a much longer wavelength, so can penetrate several cm into the food. For small items, a microwave oven can cook faster, because cooking time is limited by the time taken to penetrate into the interior. For large objects such as a whole chicken, the total energy expended is significant, and that tends to favour a conventional oven.

Humans also often like to cook themselves, or at least their skins, using shorter wavelengths, in the ultra-violet (UV). A white skinned person who tans easily will generally receive enough UV to produce a just noticeable increase in skin colouration in about 20 minutes in middle latitudes [33, see Figure 2]. Long-term tanning is correlated with skin cancer which has been heavily documented [34]. A short, 10-15 min per day exposure also provides Vitamin D, which is manufactured in the skin, but longer durations of exposure to sun does not seem to produce any more[35].

5.7 Navigation

Navigation is closely related to the measurement of distance. Determining latitude has always been straightforward: you can simply measure the solar zenith angle (angle of the sun away from the vertical), φ, and then use the relationship

$$\cos \varphi = \cos \delta \cos \vartheta + \sin \vartheta \cos \lambda \sin \delta$$

ϑ is the latitude, λ is related to the time since local noon and δ is the solar declination. $\lambda = 180t/24$ where t is the time in hours since local noon. When the sun is highest in the sky (local noon) $\lambda = 0$, and $\cos \lambda = 1$. Hence, $\cos \varphi = \cos \delta \cos \vartheta + \sin \vartheta \sin \delta = \cos(\vartheta - \delta)$. δ varies between ± 23° or so, throughout the year, so you need to know the date of the year, and then with observations of φ, ϑ can be computed from $\vartheta = \varphi + \delta$. δ is + 23° in northern midsummer 21 June, and - 23° in northern midwinter, 20 December. The procedure for finding ϑ breaks down within the Arctic and

Measuring the World, by John Austin

Antarctic circles during midwinter, when the sun doesn't rise above the horizon. Instead of the sun, observations of a different astronomical object such as the moon or a specific star, need to be made.

Unfortunately, an equivalent procedure does not exist for determining longitude. This is because any observation of astronomical bodies (sun, moon, stars) is very closely reproduced at any other longitude a fraction of a day later. There are, however, a few special exceptions due for example to the occurrence of events ,e.g., the collision of astronomical bodies, or the transit of the sun, or an eclipse. For these special occasions a specific time can be allocated. Another situation is where the position of the sun and moon can be accurately determined. The moon can then be used to obtain the longitude given a set of prior calculations. In general however, the observations of regular events such as the moon's orbit around the Earth results in what is known in mathematical jargon as 'aliasing' between longitude and time. In other words, because we are observing from a rotating Earth, an interval in time can be misinterpreted as an interval in longitude and the converse.

By the 1700s, the navigation of ships was becoming a problem because of the challenge of calculating longitude at sea. At times, ships were lost due to unexpected problems arising from navigation errors. To try to resolve this problem the British parliament in 1714 established the "longitude prize", which allocated £20,000 to anyone solving this problem[36]. The sum, after inflation would be worth several million or even tens of million pounds in today's money. Astronomical solutions were proposed, but these could not deliver the desired accuracy of 0.5 Great Circle Degrees (≈55 km) for a voyage between England and the Caribbean. One solution was to design an accurate clock that could then resolve the longitude/time aliasing. For the 0.5° accuracy, the clock would need to be accurate to 3 s per day, given a typical 6 week journey time. Such a clock needed to survive the rigours of an unsteady ship, which completely disrupts the swing of a pendulum. The first reliable marine chronometer of this sort was built by John Harrison in 1735. Harrison was something of a perfectionist and did not test his clock straight away. Instead, he worked for decades improving his clocks, before finally producing a portable watch (H-4) which was put to the test in 1762. On the journey out, the watch lost only 5 s despite 81 days at sea. On the return journey, the weather was even worse and the total error was just under 2 minutes. There was considerable infighting between Harrison and the astronomers of the time, but he was eventually awarded money in lieu of the prize[37].

An accurate clock is used in conjunction with the time of local noon to calculate longitude. The clock is first synchronised with local time. Local noon at the departure point is taken to be when the sun is highest in the sky. The clock is then set to 12.00 pm. After a period at sea, the sun is observed again, and the time, t, of local noon is determined, and compared with 12.00 pm. For a 24 hour difference, the longitude change would be $360°$, which is equivalent to $15°$ per hour of clock difference. So the new longitude is ($°W$) is $(t – 12)*360/24 = (t – 12) \times 15$. For example, if the local measurement of local noon is found to be at 2 pm on the chronometer, $t = 14$, then the longitude is $(14 – 12) \times 15 = 30$ $°W$. In principle, the sun might not be observable for many days at a stretch, so an appropriate adjustment needs to be made to the longitude taking into account of the ship velocity since the last measurement.

Navigation is the motivation for the unit nautical mile (NM), originally defined as the distance subtended by an arc of 1 minute along a great circle at the Earth's surface. The clock navigation method described above would mean that an error in the time of 1 minute would convert to 15 minutes of longitude error. At the Equator, this would mean an error in the position, assuming no error in the latitude position, of 15' of great circle arc. Taking the Earth as spherical, the position error would be 15 NM. The Earth is slightly non-spherical, so instead the nautical mile is specified as the mean value of 1852 m *(exactly)* = 607.115 ft = 1.151 miles.

The global positioning system (GPS) has now superseded all forms of navigation, providing positions to the nearest metre at least. Up to 9 satellites are used[38], and positions are determined using triangulations for microwave transmissions. Depending on the number of satellites used, positions could be accurate to a few centimetres. The timing of the signal needs to take account of atmospheric constituents (especially water vapour) as well as the curvature of space induced by the massive Earth. This shows the importance of General Relativity calculations in advanced technology applications.

5.8 Map Scaling

The drawing and interpreting of maps are important issues in society. On the global scale, the Earth's surface is projected onto a globe or a flat map. Projection onto a globe is straight forward: a fixed linear scaling will produce a fixed area scaling without any distortion of shapes of seas and continents. However, projection of a near spherical object onto a two-dimensional map can only be done approximately. The two main two-dimensional map projections are Mercator's and Mollweide's[39]. Mercator's projection keeps angles on the real Earth correct on paper, but at the cost of distorting continental shapes. Mollweide's projection is an equal area projection in which a given *area* on the Earth will map to a scaled area on paper. Other projections are possible. Typically, continents scale correctly in the tropics, but worsen towards the poles. City maps do not have these problems, as the curvature of the Earth is negligible over the size of a city.

In representing the scale of the map, the spatial units become important. 'Metric' scaling is easy to understand. Such maps can be marked as 1 in 100,000 which can be thought of as 1 *cm* = 1 *km*. Obviously, different scales are possible. Some care may need to be taken, though, as in my experience the widths of roads are not always drawn at the correct width. Instead, they are often drawn wider than expected, to ensure that they stand out sufficiently. This would mean that in some cases, road-free regions are represented smaller than reality. Maps based on the fps system usually have awkward-looking scales, such as 1:31680, which is 1 inch = 0.5 miles. I often wonder whether there is an excessive amount of precision here, though, and that on the map, 1:31680 is scarcely distinguishable from 1:32000.

5.9 Short and Long-sightedness

Corrective lenses ("glasses", "contact lenses") to correct for short (myopia) or long-sightedness (hyperopia) are prepared by the optician. Although the units are often not stated, they are already a metric unit, dioptres or m^{-1}. Even the USA uses dioptres! For myopia, the lenses have negative values or powers, corresponding to a diverging lens. For hyperopia, the lens power is positive. The lens power is *1/f* where *f* is the focal length of the lens. The theory of optics shows that the lens powers can be added[40]. Hence, two lenses of powers *p1* and *p2* have combined power

$$p = 1/f_1 + 1/f_2 = p_1 + p_2,$$

corresponding to focal lengths f_1 and f_2 respectively. To take an example suppose that you are far-sighted and that the closest focus distance that you can manage is 1.00 m. For an image and object distance *u, v* measured from your eye, which has a focal length of *f*, then the lens equation is applicable:

$$1/f = 1/u + 1/v$$

If we assume that the image distance from the eye is about 5 cm, the projection to the brain, then we set v = 0.05 m and u = 1.00 m, for the closest distance. Hence the focal length of the eye is given by

$$1/f = 1/1 + 1/0.05 = 21.$$

That is, the power of the lens eye is 21 dioptres. For the minimum distance of 25 cm, we need a focal length given by $1/f = 1/0.25 + 1/0.05 = 24$. The difference 24 - 21 = 3. So for better vision, an additional lens of 3 dioptres is needed. Of course the optician doesn't determine the power of the corrective lens by doing these calculations; instead he or she carries out some experiments until you're happy with the improved vision. The point is, though, that with a simple knowledge of lens optics you could work out what the correct lens power should be. Moreover, the lens powers are the same internationally, i.e. everyone uses dioptres, so your optical subscription is valid everywhere to everyone's benefit.

5.10 Velocity

Velocity is a combined speed and direction. Most of this section is about speed, and indeed when most people refer to velocity, they really mean speed, but velocity sounds more grand! Measuring direction is simple enough in principle, you can usually use a compass. However, that will strictly only give you the direction of magnetic north and for precise work, corrections may need to be made for true north, which would be the direction of the north pole, i.e., the axis about which the Earth rotates. The direction of magnetic north is the direction of the Earth's magnetic pole which is slightly displaced from the axis of rotation. The discussion of the Earth's magnetic field is best delayed until Chapter 12, but suffice to say that the north magnetic pole has drifted between 81 and 85 °N over the last 10 years[41]. The South magnetic pole by contrast is much further from the pole at typically 64 °S and would tend to make a compass less useful in the Southern Hemisphere.

Assuming that the direction is known (N, S, W, E etc.), perhaps measured by GPS, we concentrate now on the speed. The units of speed are m s^{-1} in the SI system, but despite its usefulness, it doesn't have a special name. The fps unit is ft s^{-1}, again without a special name. These units tend to be used for short acting processes. For example, a sprinter completing 100 m in 10 s has an average speed of 10 m s^{-1} (33 ft s^{-1}). An elevator might traverse 5 floors in a minute at an average vertical speed of perhaps 15 m in 60 s, or, 0.25 m s^{-1} (a little under 1 ft s^{-1}). More commonly, in civil use, speeds are quoted in km/h (km per hour) or mph (miles per hour). A common unit for travel by ship or air is knots (abbreviation kn, nautical miles per hour). For example an air speed of 450 kn means that an aircraft will travel 450' = 7.5° along a great circle arc in a period of one hour.

Incidentally, as a curious social comment, if you ask someone how 'far' it is from A to B, a clear request for *distance* information, in the USA you will usually get an answer in hours and minutes. The assumption is that travel is by personal car. A similar question to someone in the UK would typically receive an answer in miles.

Speeds at sea used to be measured using a log suspended from the stern of the boat[42]. Knotted rope is wound around the log and dragged through the water. The knots were placed at intervals of 14.4 m (47' 3") along the rope, and in a 28 second run time, the number of knots which unwound from the log were counted. Since 28/3600 = 47.25/6076, the number of knots in 28 s is the speed in knots relative to the current. A correction then needs to be applied for the assumed speed of current. It's no wonder that ships were lost (§ 5.7)! At least we now have GPS.

5.11 Atmospheric wind

The Extended Beaufort Scale

Force	Name	Description	Wind speed (km/h)
0	Calm	Smoke rises vertically	0
1	Light Air	Smoke drifts	1-5
2	Light Breeze	Wind felt on face; leaves rustle	6-11
3	Gentle Breeze	Leaves and small twigs move; light flags extend	12-19
4	Moderate Breeze	Small branches sway	20-28
5	Fresh Breeze	Small trees sway	29-38
6	Strong Breeze	Large branches sway	39-49
7	Moderate gale	Whole trees sway	50-61
8	Fresh gale	Twigs broken off trees	62-74
9	Strong gale	Slight damage to buildings	75-88
10	Whole gale	Trees uprooted	89-102
11	Violent Storm	Widespread damage	103-117
H1	Hurricane Force 1		119-153
H2	Hurricane Force 2		154-177
H3	Hurricane Force 3		178-208
H4	Hurricane Force 4		209-251
H5	Hurricane Force 5		≥ 252

Atmospheric wind, which of course is just the velocity of the air molecules, is an important property of the atmosphere, because of the potential damage that it can cause. It generally makes more sense to think of a wind as having the properties of the direction it has come *from*. For example a North wind is often cold and perhaps dry. In scientific work, winds are sometimes referred to as northward and this would be a wind *to* the North, or Southerly wind. Therefore some care needs to be taken in reading the terminology. Wind velocities are measured with an anemometer (speed) and wind vane (direction). The latter used to be very common as decorative pieces on church steeples and so on, but are perhaps less common in densely populated cities. Overall, westerly winds are frequent because of lower pressure towards the poles, and the application of Buys Ballot's law (§ 9.11). Wind speeds can be estimated without an anemometer to within about a factor of 2, using the 'Beaufort Scale'. This scale was invented in 1805[43] originally to provide sailors with an idea of wind speed, but was later modified to provide information so that weather reports could be compared with each other. The scale uses observations of the wind on natural phenomena to estimate a wind speed. The scale has now been extended above its original maximum to allow for hurricane-strength winds, so the description of the extended scale can overlap a bit.

   In the table, the hurricane forces are taken from the Saffir-Simpson scale[44]. Hurricanes are large tropical storms, taking their energy from the warm ocean. By contrast, tornadoes are small-scale disturbances, usually only a few hundred metres wide, and they form over the land. The path of the tornado moves at about 55 km h^{-1} or less and typically lasts just a few minutes. They are small, intense cyclones and occur typically in spring or early summer. They are well-known in the USA in 'Tornado alley' which covers Texas, Oklahoma, Kansas, Nebraska, Iowa mid-western and southern states. When conditions are ripe for their development as many as a dozen separate tornadoes can occur simultaneously, multiplying the damage they cause. For example, in Texas in May 2011, tornadoes caused over $9 Billion in damages[45]. Tornadoes also occur in Australia and Bangladesh. In all countries, they generally occur in the late afternoon or early evening. Tornado strength is measured or estimated using the Fujita Tornado Index[46], which overlaps with the hurricane scale, and is as follows: Generally, the force of the wind increases as the square of the speed, so doubling the wind speed increases the force by a factor of 4.

The Fujita Tornado Index

Force	Description	Wind Speed (km/h)
F0	Tree branches broken	64-116
F1	Trees snapped	117-180
F2	Large trees uprooted	181-253
F3	Roofs and walls torn off; cars overturned	254-332
F4	Houses levelled; cars thrown	333-418
F5	Large trees uprooted; cars thrown > 100m	419-512

Wind speed often increases with altitude, and airlines frequently take advantage of the location of strong winds, which can have a significant influence on continental travel times. The normal time between New York and London is about 7 hours, but I have crossed once in just over 5 hours, because the commercial airline was able to fly in the jet stream at 9 km altitude. On that day, the wind speed must have been about 300 km h^{-1}, and the ground speed about 1100 km h^{-1}. This journey would have saved a substantial amount of fuel, but flights from London to New York on that day were probably taking 9 hours!

As well as the wind strength, turbulence (due to variation in wind strength, especially the vertical component of the wind) is also an issue to take into account while flying. It is generally best to avoid large clouds which generate strong vertical motions around them. This can cause sudden aircraft movements in the vertical. Even clear air can have some turbulence, possibly a remnant from previous clouds that have evaporated, or other aircraft that have since moved away. Turbulence is not entirely understood, and in view of its importance to aircraft operations, particularly around airports, it is a topic of major scientific research.

5.12 Distance Measurement in Sport

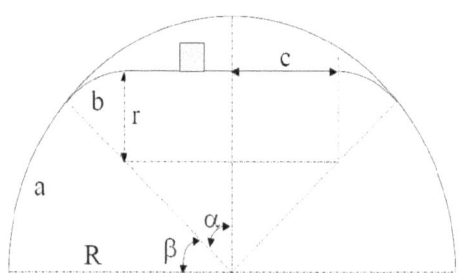

a) Calculation of the Steeplechase Lap (Water Jump inside)

		Measured	Standard IAAF	Formula
Length curve 1 (Running track):	a	.... m (+)	27.322 m (+)	$\frac{\pi \times \beta \times (R+L)}{180}$
Length curve 2 (Steeplechase):	b	.... m (+)	13.502 m (+)	$\frac{\pi \times \alpha \times (r+l)}{180}$
Length c:	c	.... m (+)	15.105 m (+)	
	z	.... m (=)	55.929 m (=)	$= a + b + c$
Steeplechase Curve:		.... m (=)	111.858 m (=)	$= z \times 2$
Normal curve:	d	.... m (+)	115.611 m (+)	
Steeplechase Curve:	e	.... m (-)	111.858 m (-)	
Shortening measure:	VM	.... m (=)	3.753 m (=)	$= d-e$
Steeplechase Lap:		.... m (=)	396.247 m	$= 400-VM$

Measurement geometry for an international running track showing the exacting requirements for its construction for use in international meetings. In this case the layout is for the steeplechase water jump[46].

Most sports have measured pitches within which the sporting activity takes place. No special measuring techniques need to be used to prepare the ground markings. Of all sports, athletics is perhaps one of the most technically demanding, and

Measuring the World, by John Austin

techniques have changed as new physics discoveries have been made. Athletics tracks for example, need to be measured to a precision of 4 cm per lap, to receive an international certificate of use[47]. That precision is 1 part in 10,000. The distance is measured 30 cm out from the curb, or for lanes 2-8, 20 cm from the outer edge of the lane marker. Lanes are 1.22 – 1.25 m wide. The track should also slope no more than 1% horizontally. These are very exacting standards and ensure that athletics world records are set at a consistently high standard.

At international level, the distances of throws and jumps are measured using triangulation and a microwave signal[48]. So, once a performance is considered valid, the measurement is displayed in a few seconds. These techniques replace the more laborious method, still used at club level, using a steel tape. Measuring this way was always a problem due to the need to keep the tape flat, not always easy if the ground has bumps due to previous implements, or indeed the activity of wild animals! Distances over 25 m are truncated to the nearest cm, supposedly the best that can be achieved. However, the real error is working out exactly where the Javelin or Discus has landed (the shot and hammer usually stay put after landing), rather than the measuring error. My guess is that at club level, the best that can be done with a steel tape is 5-10 cm, since tape errors add to the error in determining the landing position. Distances less than 25 m are also measured to the nearest cm with fractions taken down. So a jump of 8.297 m would be recorded as 8.29 m. This is much less demanding a measurement to make at club level (where in any case the jump would be more like 6.297 m!), so I think these are fairly reliable as well.

A pedometer[49] can be used to measure distances covered by foot, but it is not a high precision device, as it needs to be 'calibrated' to each different user. The instrument is attached to the waist or arm, and by sensing body motion, the number of steps is counted. To convert to a distance, the instrument needs to be calibrated by walking a set distance. This gives the stride length. Further use of the instrument relies on the stride length remaining unchanged between uses. This would clearly not be the case, for example, if the instrument were calibrated by walking, presumably a relatively short stride, and then used for jogging (an even shorter stride) or full running, which would entail a longer stride than walking.

References

[1] Ancient Roman units of measurement, Wikipedia, 27 January 2014, http://en.wikipedia.org/wiki/Ancient_Roman_units_of_measurement, accessed 27 February 2014.
[2] Ancient Egyptian Units of measurement, Wikipedia, 15 January 2014, http://en.wikipedia.org/wiki/Ancient_Egyptian_units_of_measurement, accessed 27 February 2014.
[3] Lovett, T., References for cubits from around the world, http://worldwideflood.org/ark/noahs_cubit/cubit_references.htm, accessed 27 February 2014.
[4] History of length measurement (poster), National Physics Laboratory, UK, 9 September 2013, http://www.npl.co.uk/educate-explore/factsheets/history-of-length-measurement/history-of-length-measurement-(poster), accessed 27 February 2014.
[5] Charity, M., 1995: Your body ruler – a user's manual, 5 May 1997, http://www.vendian.org/mncharity/dir3/bodyruler/, accessed 27 February 2014.
[6] Vitruvian man, Wikipedia, 19 February 2014, http://en.wikipedia.org/wiki/Vitruvian_Man, accessed 27 February 2014.
[7] Hadhazy, A., 2008: What makes Michael Phelps so good?, Scientific American, 18 August 2008, http://www.scientificamerican.com/article/what-makes-michael-phelps-so-good/, accessed 27 February 2014.
[8] League (unit), Wikipedia, 16 December 2013, http://en.wikipedia.org/wiki/League_(unit), accessed 7 March 2014.
[9] Twenty thousand leagues under the sea, Wikipedia, 28 February 2014, http://en.wikipedia.org/wiki/Twenty_Thousand_Leagues_Under_the_Sea

Measuring the World, by John Austin

accessed 7 March 2014.
[10] Metre, Wikipedia, 25 February 2014, http://en.wikipedia.org/wiki/Metre, accessed 4 March 2014.
[11] Humerfelt, S., 2010: How WGS 84 defines Earth, http://home.online.no/~sigurdhu/WGS84_Eng.html, accessed 4 March 2014.
[12] Green, T., 2007: NASA published 7 January 2005, revised 30 November 2007, http://www.nasa.gov/vision/earth/lookingatearth/earthshape.html, accessed 4 March 2014.
[13] Metre, Wikipedia, 25 February 2014, http://en.wikipedia.org/wiki/Metre, accessed 4 March 2014.
[14] Yard, Wikipedia, 18 February 2014, http://en.wikipedia.org/wiki/Yard, accessed 4 March 2014.
[15] Imperial Yard Shrinking, *The Sydney Morning Herald* (NSW, Australia), 8 June, 1950, p. 3., http://nla.gov.au/nla.news-article18159945, accessed 4 March, 2014.
[16] The BIPM and the evolution of the definition of the metre, Bureau International des Poids at Mesures, Sèvres, France, http://www1.bipm.org/en/si/history-si/evolution_metre.html, accessed 4 March 2014.
[17] Definition of the metre, Bureau International des Poids at Mesures, Sèvres, France, 11th conference Générale des Poids et Mesures (1960), http://www.bipm.org/en/CGPM/db/11/6/, accessed 4 March 2014.
[18] Definition of the metre, Bureau International des Poids at Mesures, Sèvres, France, 17th conference Générale des Poids et Mesures (1983), http://www.bipm.org/en/CGPM/db/17/1/, accessed 4 March 2014.
[19] Recommended values of standard frequencies, 633nm (I2), Bureau International des Poids at Mesures, Sèvres, France, 11 July 2013, http://www.bipm.org/en/publications/mep.html, accessed 4 March 2014.
[20] Michelson-Morley experiment, Wikipedia, 28 February 2014, http://en.wikipedia.org/wiki/Michelson%E2%80%93Morley_experiment, accessed 5 March 2014.
[21] Hawking, S., 1988: A brief history of time: from the big bang to black holes, Bantam Dell Publishing Group, 256 pp.

[22] Survey foot versus international foot: what's the difference?, Microstation Today, 7 March 2006, http://microstationtoday.com/microstation-tip-corner-survey-foot-versus-international-foot-whats-the-difference/, accessed 5 March 2014.

[23] Electromagnetic spectrum, Lawrence Berkeley Laboratory (US)http://www.lbl.gov/MicroWorlds/ALSTool/EMSpec/EMSpec2.html, accessed 5 March 2014.

[24] Newton and the color spectrum, Vision science and the emergence of modern art, WebExhibits, http://www.webexhibits.org/colorart/bh.html, accessed 5 March 2014.

[25] Flowers 'optimized' colors for bee vision, Futurity, Science and Technology, E. Walker-Monash, 7 June 2012, http://www.futurity.org/flowers-%E2%80%98optimized%E2%80%99-colors-for-bee-vision/, accessed 5 March 2014.

[26] Bird Vision, Wikipedia, 3 March 2014, http://en.wikipedia.org/wiki/Bird_vision, accessed 5 March 2014.

[27] Color vision, Wikipedia, 21 February 2014, http://en.wikipedia.org/wiki/Color_vision, accessed 5 March 2014.

[28] Ultra high frequency, 4 March 2014, http://en.wikipedia.org/wiki/Ultra_high_frequency, accessed 6 March 2014.

[29] FM broadcasting, Wikipedia, 2 March 2014, http://en.wikipedia.org/wiki/FM_broadcasting, accessed 6 March 2014.

[30] AM broadcasting, Wikipedia, 5 March 2014, http://en.wikipedia.org/wiki/AM_broadcasting, accessed 6 March 2014.

[31] Microwave oven, Wikipedia, 3 March 2014, http://en.wikipedia.org/wiki/Microwave_oven, accessed 6 March 2014.

[32] Frozen Florida, Hubpages, 22 July 2013, http://lidian.hubpages.com/hub/Frozen-In-Florida, accessed 6 March 2014.

[33] Vanicek, K. et al., 1999: UV index for the public, A guide for publication and interpretation of solar UV index forecasts for the public prepared by the working group 4 of the COST-713 Action "UVB forecasting", http://www.uv-index.ch/images/Leitfaden_COST-713.pdf, accessed 6 March 2014.

[34] Facts about sunburn and skin cancer, Skin cancer foundation, 2014,

Measuring the World, by John Austin

http://www.skincancer.org/prevention/sunburn/facts-about-sunburn-and-skin-cancer, accessed 6March 2014.
[35] Sun and Vitamin D advice given, NHS Choices, December 2010, http://www.nhs.uk/news/2010/12December/Pages/sunlight-exposure-and-vitamin-d-advice.aspx, accessed 6March 2014.
[36] Longitude prize, Wikipedia, 13 November 2013, http://en.wikipedia.org/wiki/Longitude_prize, accessed 6 March 2014.
[37] Sobel, Dava, 1995: Longitude: the true story of the lone genius who solved the greatest scientific problem of his time, Walker and company, New York books.
[38] Global positioning system, Wikipedia, 27 February 2014, http://en.wikipedia.org/wiki/Global_Positioning_System, accessed 6 March 2014.
[39] Map projection, Wikipedia, 14 February 2014, http://en.wikipedia.org/wiki/Map_projection, accessed 6 March 2014.
[40] Lens (optics), Wikipedia, 5 March 2014, http://en.wikipedia.org/wiki/Lens_(optics), accessed 6 March 2014.
[41] North magnetic pole, Wikipedia, 6 March 2014, http://en.wikipedia.org/wiki/North_Magnetic_Pole, accessed 7 March 2014.
[42] Knot (unit), Wikipedia, 3 March 2014, http://en.wikipedia.org/wiki/Knot_(unit), accessed 7 March 2014.
[43] Beaufort scale, 23 February 2014, http://en.wikipedia.org/wiki/Beaufort_scale, accessed 7March 2014.
[44] Saffir-Simpson hurricane wind scale, Wikipedia, 27 February 2014, http://en.wikipedia.org/wiki/Saffir%E2%80%93Simpson_hurricane_wind_scale, accessed 7March 2014.
[45] Extreme Weather 2011, Midwest/Southeast tornadoes May 22-27, 2011, National Oceanic and Atmospheric Administration, http://www.noaa.gov/extreme2011/joplin.html, accessed 7 March 2014.
[46] Fujita scale, Wikipedia, 16 February 2014, http://en.wikipedia.org/wiki/Fujita_scale, accessed 7 March 2014.
[47] IAAF Certification system, Report of Measurement, http://www2.iaaf.org/TheSport/Technical/Tracks/TrackMeasurementReport.pdf, accessed 7 March 2014.
[48] Mills, J., 2001: Sports Measurement, Geomatics.org.uk, http://www.m-a.org.uk/what_use/**SportsMeasurement**.doc, accessed 7

Measuring the World, by John Austin

March 2014.
[49] Pedometer, Wikipedia, 28 January 2014, http://en.wikipedia.org/wiki/Pedometer, accessed 7 March 2014.

6. Area and Volume

6.1 Introduction

Areas and volumes occur in a wide range of everyday items such as areas of paper and real estate, and volumes of drinks or the shrinking of the polar ice caps. Each of these topics requires its own set or range of units to be fully practical. Once a unit of length is defined (Chapter 5), then area and volume follow automatically by geometry. Although this is the case for the metric system, the fps system has been cobbled together using units acquired from various practices going back centuries. This has produced a confusing array of units in fps, such as the acre for area, and bushel for dry volume. Many of these units are only used in the USA and one can only hope they are abandoned in favour of the metric system sooner rather than later.

An important feature of both metric and fps is the connection between volume and mass via water. In other words, water has a density that is a round number in appropriate units: 1 g cm^{-3} or 1000 kg m^{-3} at 4 °C, its maximum density, or 10 pounds per Imperial gallon at 62 °F. The US gallon has slightly lost the connection. Some of the confusion with volumes, though, arises from the use of different units for dry volume versus liquid volume. There are pecks and bushels (dry volume) and gallons and barrels (liquid volume), but hardly anyone I've met know what these units really are, except for the gallon. By contrast, the metric system is a beacon of clarity. If few people understand them, why are fps units still in use?

This Chapter begins with a short description of units of area and volume in § 6.2. The first part of the Chapter covers area and the second part volume. § 6.3 – 6.5 covers small, medium and large items – microscopic, human-sized and land-sized respectively. § 6.6 suggests approximate area conversions between unit systems. § 6.7 addresses volume measurements, with emphasis on liquids and gases (fluids). The metric system uses the same units for liquid and volume but the Imperial system does not, and so in § 6.8 we discuss dry volume measurement. § 6.9 exposes the many peculiar units still in use in some places and § 6.10 gives approximate conversions for those who must convert. Finally, in § 6.11 we discuss a serious issue of the current time – the Earth's changing icecaps.

6.2 Units of Area and Volume

The main units of area in the metric system are cm^2, m^2, and km^2, while in the fps system the units are in^2, ft^2, $yard^2$ and $mile^2$. Conversion factors are given in Appendix B, while the 'rule of thumb' or approximate conversions are given in §6.6.

The main units of volume in the metric system are cm^3, m^3, and (rarer) km^3. Litres, containing 1000 cm^3, are also frequently used. Although often abbreviated as a lower case l, in most typescripts this is difficult to distinguish from the number 1, and letter I, and so the upper case abbreviation L is now more commonly used for litre and I will follow this practice. In the fps system in^3, ft^3, $yard^3$ and $mile^3$ all exist, but there is often a distinction between dry volume and liquid volume, a distinction which does not exist either in physics or the metric system. Moreover, unlike the metric system, there is no clear connection between bushels, gallons, barrels etc. and cubic inches or feet, at least no connection that this author has been able to find.

6.3 The Areas of Small Items

Naturally, the unit depends on how small is 'small', but I am thinking here of items with linear dimension of about 1 cm or less. The natural units to use then are mm^2 or even μm^2 for the microscope. When expressed like this it is assumed that the full unit mm or μm is squared (see Chapter 2). These units don't really have effective equivalents in the fps system. Areas can be given in square inches of course, but much smaller areas tend to be the province of specialist fields of interest, rather than everyday experience. Specialist fields are dominated by units which are typically metric. The themes of the very small and very large are developed further in Chapter 15, but the same concept applies: if the measurement is outside our direct experience measurements tend to be exclusively metric. Those using the fps system therefore tend to use imperial units at intermediate sizes, but switch to metric once personal intuition is of no help (either because items are very small or very large). By contrast, those who use the metric system just switch to a new power of 10 once a given unit is inadequate.

6.4 The Area of Medium-Sized Items: Paper Magic

Here, I am considering medium-sized items to have a linear dimension of about 1 m. In metric, the floor area of a house is usually quoted in m^2, while in the fps system, sq. ft. is typical. Of course the latter is more difficult to calculate, as the inches have to be converted to feet, as a decimal for example. Paper sizing for office use could also be given in m^2, but office work consumes so much time and effort that the metric system adopted its own paper sizing, the A notation. There is also a B and C notation under International Organisation for Standardization (ISO) 216[2], which is sort of strange as being international, the USA doesn't seem to

Measuring the World, by John Austin

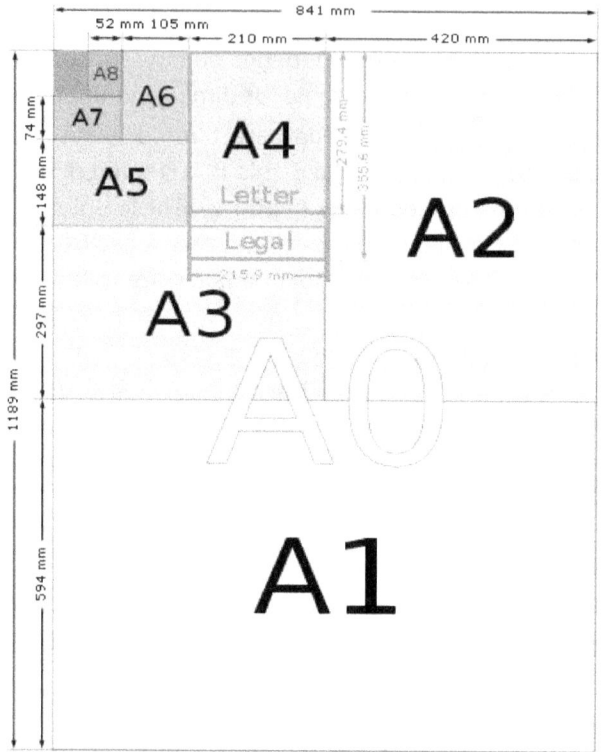

A Series paper sizes, image courtesy of Office 365[1].

have any interest in it, yet the spelling of the organisation name seems to be US, with a 'z'. Anyway, I call the paper sizing paper magic, as I think it represents one of the most elegant ways imaginable for sizing and comparing paper. Here are the rules:

1. The area of paper sheet An is 2^{-n} m^2. Thus A0 is 1 m^2, A1 is 0.5 m^2, A2 is 0.25 m^2..... A4 (a fairly heavily used paper size) has area 1/16 m^2.

2. The ratio between the long (y) and short (x) dimensions of

the rectangular paper is $y/x = \sqrt{2}$.

The beauty of the paper sizing is that standard A pieces of paper can be placed on top of each other without overlap due to the $\sqrt{2}$ aspect ratio[3]. Starting with A0 in portrait mode, it can be covered in two sheets exactly of A1 paper in landscape mode. In turn, each of the A1 landscape sheets can be covered exactly by two A2 sheets in portrait mode. Hence, progressing to smaller paper successively switches mode from portrait to landscape to cover the previous size paper exactly. If A4 is too large for the needs of the user, then A5 or A6 paper may be used instead. By standardising all paper sizes, related products such as files and folders can also be standardised. Utility companies, to choose one example, typically have standardised their paper bills at A5. I know of no similar standardisation in relation to imperial-sized paper. Foolscap (approx. A4) used to exist in Britain, but it has been dropped in favour of A4. The USA still routinely uses odd-sized paper and utility bills have their own unfathomable rules. Legal paper is also awkwardly shaped: the same width as ordinary paper, but a few inches longer. As a result it doesn't fit into a photocopying machine, which is a nuisance for anyone who has bought a house in the USA!

6.5 Measuring Large Items – Land Area

A unit of measurement in the metric system is the are = 100 m^2. It is more easily recognised with the hecto multiplier: 1 hectare (1 ha) = 100 are = 10^4 m^2. I am not completely happy with this unit, although I accept that it has its adherents. My main objections are that the are is almost never seen on its own, so the relationship to the base unit of m^2 is lost on most people. So even though most people would be able to visualise, say 100 m^2, rather fewer people I think understand how large 1 ha is, except that they know its a few acres. However, as we see below few fps adherents know how big an acre is either. Secondly, both the unit (a) and the common (one would be tempted to say only pre-multiplier are not factors of 10^3 as recommended by SI. It does have its advantage, though, in working with areas that it is useful to visualise the amount of space by taking the square root. Thus 1 are = 10 m x 10 m, approximately the size of a UK suburban garden (like mine!). Similarly, 1 ha is 100 m x 100 m, about the size of two football pitches. Once areas become much larger than this, it is probably better to use km^2. So in practice, ha tend to be used almost exclusively by Estate Agents, at least the less conservative of them (in Britain). The more conservative tend still to use acres, which is a weird unit if ever there was one!

In English-speaking countries, land area is traditionally measured in acres, which is a unit from the fps system. 1 acre is defined as 4840 square yards. You what? Yes, exactly, but it is not as bad as it looks. The arithmetic involved to convert units in and out of acres provides a headache for the unwary, but there are some tricks to ease the computational burden. An acre (abbreviation ac) is the area that traditionally could be ploughed by an ox without taking a rest, although there seems to be some uncertainty with Wikipedia indicating that more than one ox is involved, and in the middle ages, they apparently took all day[4]. Presumably the fields were firmer in those days. An acre is a chain wide (22 yd) and a furlong (220 yd) long. So even though areas in square yards are awkward, if distances can be broken down into chains and furlongs, calculations can be performed in the head. An acre is also 4 by 40 rods, where of course as we have seen before, 1 rod is 5.5 yd exactly. Also the furlong was considered to be the length of a furrow in English farming, although in practice, the length varied. This relationship to the farming community has long since disappeared, following mechanisation of activities, but the old units stubbornly remain. To determine the number of acres in a square mile, you can visualise the area as 80 chains by 8 furlongs giving 640 acres per square mile. The number 640 hardly rolls off the tongue, but somehow it seems more palatable than working directly in square yards: 1760 x 1760/4840 = 640. You need to be pretty proficient with numbers to see that in your head! Generally, converting from square yards to acres or the converse can be a lot more difficult. So why do people still do it? In the metric system, km^2 is an easy unit to work with and it is easy to transform to ha if needed: 1 km^2 = 10^6 m2 = 10^2 x 10^4 m^2 = 10^2 ha = 100 ha. Easy, peasy!

6.6 Approximate Conversions – Area

Similar conversions between units of area can be used as described for length (Chapter 5). For example, the conversion from sq. yd. To m^2, is a factor 0.9^2 = 0.81. Hence, a reduction of 20% would give a good approximation (error 2%). To convert from m^2 to yd^2, divide by 0.8 or multiply by 1.25; in other words add 25% (error also 2%). Sometimes the dimensions might be given separately and it may be better to convert each dimension separately. For example, a field 150 yd x 350 yd is approximately 135 m x 315 m = 42000 m^2 = 4.2 ha = 10.5 acres approx. This uses the approximation 20 m x 200 m ≈ 1 acre. 20 m x 200 m = 0.4 x 10^4 m^2 = 0.4 ha. So, 1 ha is about 2.5 acres. Note the power of approximation here: sometimes it is easier to use the metric system as a means of making an easier calculation in fps: convert first to metric, do the calculation in metric, then convert back to fps. This theme comes up from time to time in converting units, not just area ones: if an approximate calculation (error of a few %) is good enough then the mental arithmetic is invariably easier in metric, counterbalancing the additional effort required by two conversions of units.

Other approximations include 1 ft^2 = 0.3048^2 m^2 =

0.092903 ≈ 0.1 m². So a reasonable approximation is to take 1 ft² ≈ 0.1 m² or 1 m² ≈ 10 ft² and this gives an error of only 7%. This may be good enough approximations to be able to appreciate floor areas in buildings.

A fun question to ask is: how many regular sheets of paper are needed to cover an acre? Using A4 paper, we have 2^4 = 16 sheets m^{-2} and an acre is about 20 m x 200 m = 4×10^3 m². So the number of sheets is about $16 \times 4 \times 10^3$ = 64,000, a calculation that can be done in the head. Try that in the fps system without a calculator!

6.7 Volume Measurements of Fluids

The metric system provides a flexible range of units from which the appropriate sized unit can be chosen: mm³ (drug volume in a syringe), cm³ (also referred to as mL – millilitres), L (along with mL size commonly used for drink volume and every day use), m³ (volumes of water usage or natural gas use for domestic power). These units are all related to each other of course by a factor of 1000 in volume. The litre is one of the most useful measures, which is why it has a separate name. Petrol is sold by the litre and bottles of drink from the supermarket or off-licence are often sold in 1 or 2 L bottles or containers. Wine is sold in 0.75 L bottles but for some strange reason, other alcoholic drinks are sold in 0.7 L bottles. A litre is the order of size of your stomach so a 1 L drink will probably necessitate a trip to the bathroom. The density of water is almost exactly 1 g cm^{-3} or 1 kg L^{-1} or 1000 kg m^{-3} at its maximum density of 4 °C. Most liquids (except molten metal) are about the same density of water, so a rough and ready conversion to weight exists if you know the volume. For example, that 2 L bottle of soda will weigh 2 kg as it comes in a very light plastic bottle. That will give you an idea as to whether it can be

Measuring the World, by John Austin

carried in a flimsy plastic bag!

Measuring jug used for domestic purposes. From Argos catalogue[5].

In the fps system of units, gas and liquid flows are often measured in cubic feet, which is smaller than 1 m^3. A smaller unit still is the fluid ounce (fl oz), which is 28.413 mL in the UK (where used) and 29.574 mL in the USA. 1 fl oz of water at 62 °F has the useful property of weighing almost exactly 1 oz (28.35 g) in the UK. The original intention of the metric system was to link spatial and mass dimensions through water at its maximum density, but in fact modern measurements give the density of water of 999.972 kg m^{-3}[6], rather than 1000 expected, but the difference is negligible for most purposes. However, there are slight knock-on effects in the size of the fluid ounce in mL being slightly different from the size of the ounce in grammes. The UK gallon has 8 pints (4 quarts) of 20 fl oz each. Thus the UK pint is 568.26 mL and a gallon of water is 4546.09 mL = 4.54609 L. Interpolating between the densities of 15 °C and 20 °C, the density of water at 62 °F is about 998.80 kg m^{-3}, so a gallon of water at 62 °F weighs 4.5406 kg almost exactly 10 lb (an error of only 0.1%). In terms of linear measurements, the UK gallon is also 277.419 cubic inches. The US gallon also has 8 pints, but they are smaller – 16 fl oz (not 20) of 29.574 mL = 473.2 mL = 231 cu in exactly. This gives a total of 3.7855 L. So, the UK gallon is larger by 20.1%. If the US fl oz were designed with a specific liquid in mind, 1 oz would have a volume 29.574 mL. Its density is therefore 28.35 g/29.574 cm^3 = 0.9586 g cm^{-3}. The minimum density of water is this value just below the boiling point (958.4 kg m^{-3} at 100 °C[6]), which doesn't seem a useful liquid, and the only other liquid that has the right density is castor oil[7]. In other words, the elegant link between linear and weight measurements, present in the UK fps units and the metric system, break down slightly in the US system. In the UK, when the fps system was more extensively used, pints tended to be the preferred volume measure. In the USA, quarts are more commonly used, and these are quite similar in size to the litre (1 US quart = 0.946 L).

6.8 Dry Volume Measurement

The metric system does not distinguish between liquid and dry measures, for good reason – namely, there is no fundamental reason to treat them differently and to do so is unnecessarily complex and confusing. So the same units described in § 6.7 are also used for dry measure: m^3 for large volumes, L for smaller ones.

By contrast, the fps system revels in confusion and complication, and has another set of units just for dry measures, dating back to the time, long since gone, when it was useful to

Measuring the World, by John Austin

distinguish between them. There are still cups, pints and quarts (but not fl oz!) but the dry volume is 16% larger than the liquid volume! The argument is that the extra volume was added to allow for trimming of vegetables and so on to remove the inedible parts[8], but this is obviously a rough and ready measure. Apparently, when measuring cooking ingredients in the USA, 'liquid' measurement is also used for ingredients such as flour, sugar, shortening, butter and spices[9]. This is a new meaning of the word 'liquid' that I have never come across before! The sad thing is that although the differences between dry and liquid measurement exist in the USA, the public seem oblivious to it all. The main measure is the quart (67.2 cu in) at 1.1 L, but several silly units are introduced (§ 6.9). Amazingly, some of these are still actually used.

I have tried to do some research into the cost of cinema popcorn, and failed miserably. This is of course what the market will bear and that's as it should be. It appears that many people complain about the cost of popcorn in the cinemas, as something like a factor of 10 or more markup on the cost of making it[10]. Personally, I don't care: the cinema could make the price £1000 per box. As in anything, if you don't like the price, nobody forces you to buy it. What concerns me more is that I have failed to understand how it is priced. It appears that the basic popping corn is bought by the retailer by weight and sold by volume to the consumer. But how big is a small, medium or large box? In the UK there is probably no confusion it is just sold in a box, which to me looks like about a litre for a medium box. A litre is a litre is a litre. In the USA, though it might be sold as a quart and then what is it? Is it a liquid quart or a dry measure quart? Do the relevant standards authorities have regulations in place for the selling of popcorn?

Measuring the World, by John Austin

6.9 Peculiar Units

What is a peck or bushel? Whenever I have asked this of my American friends I have usually received a blank face, and these units are in relatively common use there! Residents of other countries fare no better, but at least they are not in common use. And who can blame their ignorance? How many people know how big a cubit is, after all? However, peck and bushel should also be confined to the dustbin of history. It is even difficult to find someone who knows what sort of units these are (length, volume, mass?). Yet occasionally I have bought a bag from the supermarket in the USA containing 1/2 peck of apples. Here is a summary: 1 peck = 8 (dry) quarts = 8.81 L and 1 bushel = 4 pecks = 35.239 L. The bushel was originally equal in volume to a cylinder 18.5 in diameter by 8 in high which is not an exact number = $\pi r^2 h$ = 3.14159...x9.25x9.25x8 ≈ 2150.42 cu in, which has been rounded to 2150.42, 1 part per 10 million less than the exact number, close enough for most purposes! The bushel (abbreviation bu) is frequently used to measure corn volume and its price is very important to the US economy. Yet it seems that very few people know what it is. In less enlightened times, Britain also used bushels, although they were smaller in size. Currently, although commodity values are often per bushel, they are more normally sold by mass, implying a specific density[11]. For example, one bushel of wheat is sometimes taken to weigh 27.2 kg.

The 3rd poorly known unit is the barrel (abbreviation bbl), as in barrel of oil. If the price of a bushel of corn is important, the price of a barrel of oil is doubly so! Yet again very few people have a precise idea of what it is. While most people would guess correctly that it is a volume measure, very few indeed would know that it is equal to 42 US gallons = 158.991 L[12]. Why 42, anyway? Except perhaps it's the answer to life, the universe and everything! Without a proper set of unit pre multipliers in place in the fps system, each industry is free to choose its own. Mbbl is the abbreviation used by US oil companies to mean thousands of barrels, and MMbbl is used to denote millions. The first b in bbl can sometimes be confused for billion especially when used in financial circles. It does concern me that these units are continually used, despite very few people knowing what they are. Just to add to the confusion, a barrel of beer is 143.85 L and a barrel of wine 119.24 L!

Several other units are in use in the building trade. The cord is 128 cu. ft., originally used for measuring firewood and the yard is sometimes used to mean 1 cubic yard. The yard (cu. yd.) is also sometimes called a 'load'. At least the metric system has some uniformity. Apparently, the 'cord' is not recognised by the US government, but I'm not sure what value this has, bearing in mind that the metric system *is* supposed to be recognised, but also isn't used.

Apparently, the annual total water use in the state of Texas is 18 million acre-feet[13]! Somebody should be really ashamed. Whatever is wrong with calling this 22.2×10^9 m^3, or 22.2 km^3, if you prefer?

6.10 Approximate Conversions – Volume

The existence of numerous US units is a potential source of major confusion. To avoid adding to that confusion, I give approximate conversions for just the major units.

A good approximation for the US (liquid) quart is one litre, or more accurately, subtract 5% from the volume in quarts to obtain the volume in litres (0.4% error). To convert UK pints to litres, double the number and subtract 10% (error 2%).

To convert from US gallons to litres multiply by 4 and subtract 5% (error 0.4%). For UK gallons to litres, multiply by 4 and add 10% (error 3%). Reverse the calculations to go from litres to gallons: divide by 4 and add 5% (US, error 0.6%), or divide by 4 and subtract 10% (UK, error 2%). A barrel of oil is almost 160 litres. To convert approximately from barrels to litres multiply by 160 (error 0.7%) or 150 (error 6.6%). To convert approximately from litres to barrels multiply by 6 and divide by 1000 (error 4.7%).

For dry measures, 1 peck = 9 L approx. (2% error) and 1 bushel = 35 L (1% error). To convert from litres to pecks or bushels divide by 9 and 35 respectively.

6.11 The Earth's Changing Ice Caps

Over the last few decades, large changes have been seen in the area and volume of Arctic ice which have been disproportionate to other changes in climate[14]. In the Antarctic, ironically, the opposite effects have been seen, with climate change melting continental ice and the resulting cold fresh water flows have insulated the sea from the land allowing more ice to form[15]. The Arctic is considered serious enough to be reported in the mainstream media, which does not always get its conversion factors right. For 2012, the ice area minimum was 3.41×10^6 km^2 on 17th September[16] and the conversion from sq. km to sq. miles is to divide by 2.58 (Wall Street Journal take note!). The previous minimum in 2007 was 4.07×10^6 km^2, possibly paving the way for an ice-free summer within 40 years. The volume of ice in the summer is also at record low levels, although the area tends to be a more emotive issue. Nonetheless, the volume is falling faster than the area because multi-year layers of ice are not being established.

References

[1] Paper, Office 365, https://www.office365.co.uk/1/Paper/1, accessed 3 March 2014.

Measuring the World, by John Austin

[2] Paper size, Wikipedia, 4 March 2014, http://en.wikipedia.org/wiki/Paper_size, accessed 7 March 2014.
[3] Lichtenberg's letter to Johann Beckman, written in 1786, http://www.cl.cam.ac.uk/~mgk25/lichtenberg-letter.html, accessed 7 March 2014.
[4] Acre, Wikipedia, 27 February 2014, http://en.wikipedia.org/wiki/Acre, accessed 7 March 2014.
[5] Measuring jug, Argos, http://www.argos.co.uk/static/Search/searchTerms/MEASURING+JUG.htm, accessed 7 March 2014.
[6] Properties of water, Wikipedia, 24 February 2014, http://en.wikipedia.org/wiki/Properties_of_water, accessed 8 March 2014.
[7] Liquids – densities, The engineering toolbox, http://www.engineeringtoolbox.com/liquids-densities-d_743.html, accessed 8 March 2014.
[8] Dry measure, Wikipedia, 13 November 2013, http://en.wikipedia.org/wiki/Dry_measure, accessed 8 March 2014.
[9] Goodcooking cooking measure converter, Good Cooking, January 2014, http://www.goodcooking.com/conversions/liq_dry.htm, accessed 8 March 2014.
[10] "Popcorn and price discrimination", by Geoff Riley, 24 February, 2008, http://www.tutor2u.net/blog/index.php/economics/comments/popcorn-and-price-discrimination, accessed 8 March 2014.
[11] Bushel, Wikipedia, 20 February 2014, http://en.wikipedia.org/wiki/Bushel, accessed 8 March 2014.
[12] Barrel (unit), Wikipedia, 7 March 2014, http://en.wikipedia.org/wiki/Barrel_(unit), accessed 8 March 2014.
[13] J.-P. Nicot, 2012: Hydraulic fracturing and water resources: a Texas study, Abstract, Annual meeting & exposition, Geological Society of America, 4-7 November 2012, Charlotte, North Carolina, USA, https://gsa.confex.com/gsa/2012AM/webprogram/Paper207059.html, accessed 8 March 2014.
[14] Sea ice, Arctic Report Card: Update for 2013, National Oceanic and Atmospheric Administration, 17 December 2013, http://www.arctic.noaa.gov/reportcard/sea_ice.html, accessed 8 March 2014.

[15] Bintanja, R. et al., 2013: Important role for ocean warming and increased ice-shelf melt in Antarctic sea-ice expansion, Nature Geoscience,
6, 376-379, doi:10.1038/ngeo1767, published 31 March 2013.
[16] Data from the National Snow and Ice Data Center, Boulder, CO, USA, http://nsidc.org/arcticseaicenews/, accessed 8 March 2014.

7. Mass

Mass - the measure of inertia of an object - its tendency to continue at constant velocity or its resistance to acceleration

7.1 Introduction

Mass is often colloquially referred to as 'weight' and is one of the base units in any system. Mass is defined above, whereas weight is more correctly the force of a mass under gravity (§ 7.3). As noted in Chapter 2, mass is the only base unit which is not currently held as a universal standard. These are extremely exciting times for physics and metrology, the science of measurement, as these sciences have developed to the point where such a universal standard will likely be realised within a few years, as described in Chapter 17. Once the experimental precision for the kilogramme primary standard is improved further, and the physics behind the experiments confirmed to be valid universally, the current prototype can be replaced. For the present, the prototype is a cylinder of platinum iridium alloy and is protected from environmental damage by a sequence of laboratory and security measures. Moving to a universal standard would remove the need for the prototype and associated costs and make the standard much more readily available to any laboratory in the world. More importantly, the move to a universal standard would protect against further environmental damage of the prototype which has been observed over the last century despite substantial efforts to protect the prototype[1].

Measuring the World, by John Austin

The metric system uses water as an important link between length and mass units, as noted in Chapter 6. Water is one of the most valuable commodities on Earth because of its life-supporting properties. Water is also one of the most plentiful: there is about 1.4×10^9 km^3 of it, of which 3% is fresh, and 97% of it in the form of sea water[2]. Of the fresh water, three quarters is in the form of ice caps and glaciers. Water is fairly inert chemically, so it tends *not* to be used, but merely 'borrowed'. Nonetheless industry 'borrows' large amounts of water, often returning it to the land with dissolved impurities (pollution). A single Sunday newspaper needs 300 L of water to be produced, while steel making requires almost 30 L kg^{-1}[4]. Environmental disasters occur not infrequently due to an excess (floods, land slides) or insufficient fresh water (drought, forest fires). The importance of water is built into the metric system: 1 m^3 of water at 4 °C has a mass of 1000 kg (the precise maximum density is 999.975 kg m^{-3} at 3.984 °C[5]). This means that a rough and ready set of units can be set up by any isolated community, based approximately on SI units. Such rough and ready units could compare well with the primary standards!

The fps system also relates volume to mass units. For example 1 UK gallon of water weighs 10 pounds at 62 °F (Chapter 6). However, the relationship between linear dimension (feet) and gallon is not as clear cut as in the metric system. In the US system of units, as noted in Chapter 6, 1 gallon of water weighs 8 pounds, but only close to its minimum liquid density of 958 kg m^{-3} at 100 °C. This is a less practical relationship, but assuming this relation holds for lower temperatures may be good enough for some work.

Mass could be thought of as an amount of substance, but another way of looking at the 'amount of substance' is in terms of the numbers of atoms or molecules. This is a very useful concept for chemistry in general and gases in particular, and is discussed in detail in Chapter 13.

Measuring the World, by John Austin

In § 7.2 the history of mass units is described, and in § 7.3, the difference between mass and weight is explained. Comparisons between units is made in § 7.4, and some peculiar units are mentioned in § 7.5. Weighing machines have a rich history and are discussed in § 7.6. In § 7.7 situations in which weight matters and when it doesn't are explored. Density and specific gravity are derived from mass and are discussed in § 7.8. Simple scaling which uses some of these ideas is discussed in § 7.9. In particular, we see that the comparisons of the strengths of humans versus some other animals, on the basis of the size or weight can be quite misleading.

7.2 The History of Units

Once the need for standards became clear, for example to form the basis for commerce, it was necessary to establish a unit of weight. Many of our early units derive from the Romans who gave us the pound[6] (libra in Latin) hence lb abbreviation, although it was about three quarters the size of the current pound, with only 12 uncia (Roman ounces) per libra. By the middle ages the weight became the Avoirdupois (to have some pears!) unit of France[7]. The kilogramme came into existence in the 1600s. Over the centuries since, the units were made more rigorous by ultimately creating artefacts of a metallic alloy to serve as the primary standard against which all other masses were compared. The pound was kept in the standard department of the Board of Trade[8] and the kilogramme is maintained at the International Bureau of Weights and Measures (BIPM) in Sèvres near Paris [9]. In the interests of global agreement, the Avoirdupois pound was abandoned in 1959 and redefined as 453.59237 g exactly. The international prototype kilogramme is housed within a sequence of three bell jars to minimise atmospheric contamination, but as noted previously, this has not prevented a drift in its mass, or of that of the international copies over the last century[1].

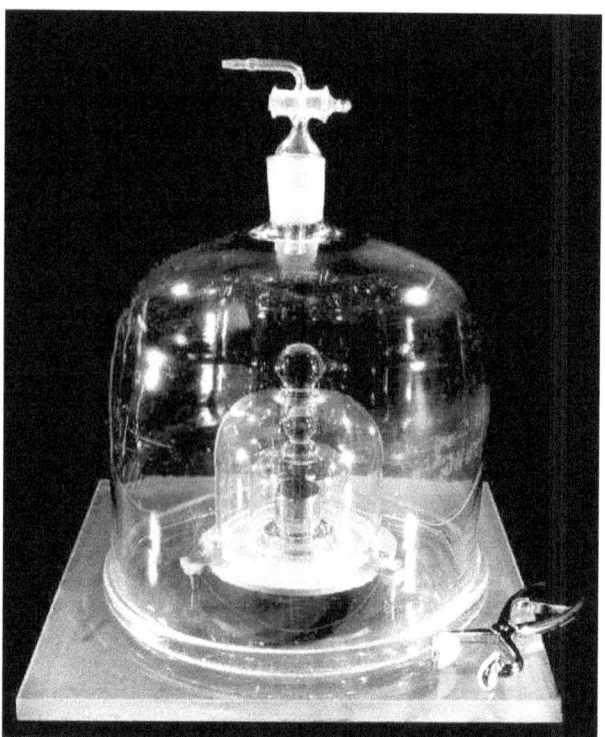

The prototype kilogramme, held at the BIPM[9].

7.3 The difference between weight and mass

The two terms weight and mass are often used synonymously by the public and even in technical work precise wording is not always adhered to. Weight is properly used to define the force on an object under gravity. For example, in orbit around the Earth, astronauts are weightless (they have zero weight) whereas their mass of course has not changed noticeably since leaving Earth. On the moon where the gravity is 1.6249 m s^{-2}[10] (1/6 of that on Earth), the astronauts weights are also 1/6 of the weight on Earth. Weight is the product of mass and acceleration due to gravity (recall Newton's law $F = ma$, where F is the force, m is the mass and a is its acceleration). For objects on the Earth's surface, $a = g = 9.8$ m s^{-2} = 32.2 ft s^{-2}. The weight should strictly be measured in newtons (N). I'm 65 kg, so my 'weight' is 637 N. In the fps system, the unit of force is the poundal, but confusingly pounds are sometimes used for forces. My weight in fps is, strictly, 143 g poundals = 143 x 32.2 = 4580 poundals, a completely unrecognisable figure! If someone asked my weight, I would be correct in saying 637 N, but it would confuse most people who would better understand 65 kg. So, when people ask my weight, they really want to know my mass. Of course I know that, and I don't stand on ceremony. An amusing (at least if you are an academic) issue related to this, is how can you easily lose weight? Well, the answer is to travel to the Equator where g is a few tenths % smaller than at the poles (9.78 v. 9.83 m s^{-2}) primarily due to the Earth's rotation. Whether anyone would be prepared to undergo such a journey for a few newtons weight loss that is only temporary, I don't know. If fad diets are anything to go by, I think they would!

7.4 Comparison between units

Where the fps units currently persist, there is a tendency to use different versions. For example, in the USA, units are pounds (453.59237 g) short tons (2000 lb, 907.18474 kg) and long tons (2240 lb, 1016.0469088 kg), with short tons more common and simply called 'tons'. Sometimes in the US, the word ton is used to mean 'long ton', so one is sometimes left guessing. In the UK, which is of course officially metric, pounds and stones (14 lb) are often used unofficially to weigh people. There are 8 stones in a 'hundredweight' (=112 lb) and 20 hundredweight in a ton. In the UK, when ton is used, it almost invariably refers to long tons of 2240 lb. A hundredweight (cwt) is a useful unit for weighing vehicles and their loads, but it's not as useful as the tonne (1000 kg)!

In accordance with SI recommendations, commonly used units are multipliers of a power of 10^3 above and below the base unit (g). Tonnes (t) have their special name in recognition of the usefulness of the unit in commerce. Both large and small units are easily accommodated by the metric system.

For very small units, the fps system is less flexible. There is the dram (1/16 oz), popular amongst Scottish whisky drinkers, but their 'wee dram' is somewhat larger than 1/16 oz, one supposes. A smaller unit is the grain = 1/7000 lb = 64.79891 mg. This is a very odd sized unit but it has some colourful history[7]. These units are not used very much any more. In the USA it is not uncommon to mix units with ounces and grammes together, for example 5 oz 15 g.

A very good approximation (error 0.2 %) in calculating pounds from kilogrammes is to double the number and add 10%. To convert from pounds to kilogrammes the inverse operation (Subtract 10% and halve) yields an error of about 1%. Another very good approximation is to take 1 (long) ton = 1 tonne. This gives an error of only 1.6%, with the ton slightly larger. In many cases, commercial data may not be accurate to better than 10% anyway, so this will be a very good approximation. The similarity of the words ton and tonne stimulates the press to call the latter 'metric tons'. While this may be fine until the audience becomes used to the word tonne, I really don't think the audience pays any attention. So the press seem to be wedded to using twice the space when space is always at a premium.

Measuring the World, by John Austin

7.5 Peculiar Units

The 'troy' system of weights is believed to have been developed in Troyes (France) in about 1300. Amazingly, these units are still used today to weigh precious metals, coins and jewels. In this system, 1 Troy pound is 5760 grains, compared with 7000 grains Avoirdupois. I Troy ounce, though, is 480 grains, so there are 12 Troy *oz* to the troy pound. The troy ounce is further divided into 20 pennyweights, or 1 pennyweight = 24 grains. It does suggest (numbers such as 12, 20, 24, and 5760) that a lot of imagination has been put into these systems, but not much thought.

US engineers like to use their own set of special units, as we have found out previously, and of course this is a common cause of miscommunication with the rest of the world. I suppose it makes them feel superior. Chapter 16 shows how on more than one occasion, they have managed to confuse themselves as well! They sometimes use a mass unit of 1 slug[11]: "a force of 1 pound will accelerate an object of 1 slug by 1 ft s^{-2}". This is a messy definition as a pound isn't a force (poundal is the fps force – Chapter 9), but the implication is the slug is *g* pounds of mass = 32.17405 lb, based on standard gravity. A fixed value of *g* is involved, otherwise the slug would vary from Equator to pole. That is the problem with some of the fps-derived units – they are not always rigorously documented.

As if this were not enough, there is also a unit for measuring the mass of precious stones, the carat, which is defined as 0.2 g[12]. Only 'common' people would have defined it as 0.1 g, *a* more obvious amount. Anyway, I'm completely mystified as to why the gramme is not good enough. Certainly carats are confused with carats[13] (same spelling UK, but spelt karat in the USA) which is another unnecessary unit indicating the purity of precious metal. 18-carat gold is 18/24 pure gold, but why this can't be described as 67% pure I don't know. Perhaps the industry likes to cover up the relative impurity of some trinkets, by shrouding the language in jargon.

7.6 Weighing Machines

Two general types of weighing machines are in general use. The first works on the lever principle. The steelyard, which is the precursor to the modern 'medical' scales, was used by the Ancient Romans 2000 years ago. A pan or hook was used to hold the heavier load, attached to the short arm of the device. A small weight was moved along the long arm until balance was reached. The markings on the indicate the mass calculated by the lever principle. A similar type is the balance for measuring small objects. It has two pans with the same length arms on both sides. The balance can weigh objects to a precision of about 0.1 mg with care. The image below shows an antique analytical balance, which enables us to see the workings of the device. There are more sensitive balances for scientific work, which uses a single pan balanced by internal weights. The pan holding the sample is usually protected in its own compartment, otherwise the apparent weight of the sample would oscillate wildly. These machines are useful for measuring small amounts of chemicals, fractions of a gramme, and may have a precision of 1 µg[15]. These balances all work on the principle of calculating the torque (weight x length from the axis) and give an absolute measure of mass since the comparison weight is subjected to the same gravity. However, the machines would not work in space, in a gravity-free environment, although they would give the correct mass on the moon.

Bathroom scales in people's homes generally work on the spring or strain gauge principle. A weight on the scales compresses a spring. The heavier the weight, the larger the compression. The scales need to be calibrated for known weights. That is, a known weight compresses the spring a fixed amount.

Measuring the World, by John Austin

Antique analytical balance in mahogany cabinet. Image from Wikimedia[14].

Weighing scales for customers to weigh produce in grocery shops are also typically a spring type, but it is sometimes arranged such that the weight stretches the spring. The more the spring is stretched, the heavier the weight. In all cases the spring assumes standard gravity and in principle an incorrect mass would be measured if the local gravity differed from the standard amount (taken to be 9.80665 m s^{-2}).

Lever machines correctly measure mass, as they compare the force under the same gravity of two separate objects. They would give an accurate reading of *mass* on the moon at 1/6 gravity. By contrast, the spring machines correctly measure *weight*, so on the moon the weight would be 1/6 that of the same object on the surface of the Earth. To measure mass at zero gravity would require the measurement of the gravitational force between two objects, although this is extremely small.

Measuring the World, by John Austin

7.7 When Your Weight Matters and When it Doesn't

The increasing incidence of obesity rates in the developed world has prompted the invention of the Body Mass Index (BMI) which is mass/height^2 in SI units[16]. For me, this works at $65/1.72^2 = 22.0$ kg m^{-2}, close to the middle part of the 'normal' range (18.5 – 24.9). BMI over 25 is considered 'overweight', and BMI over 30 is considered 'obese'. About 30% of adults in the USA and about 25% in Britain and Europe have BMI over 30, and an increasing number of children are showing high BMI in the developed world. The description 'obese' is an oversimplification of course, since a well-muscled individual could score highly on BMI, but the excess weight is not fat. The BMI is just one of a number of health measures. In fps units, to reproduce the same BMI numbers, a multiplicative constant is needed: BMI = 703 M_f/h_f^2, where M_f is weight in pounds, and h_f is height in inches. At least the BMI itself is standardised across all countries, even though it may be obtained by different routes.

An example where an individual's weight doesn't matter is in the calculation of aircraft loads to determine fuel requirements. Have you ever wondered why airlines weigh your luggage, but not you? There have been threats to do so, but the reason for this has stemmed from the need to maximise revenue rather than the need to calculate fuel requirements more accurately.

A person's weight is approximately normally distributed about a mean value of about 70 kg[17, taking US values as more representative of those who actually fly]. The normal distribution is very common in nature and describes how in a given population, some people will be heavier than average, and some will be lighter. The variation can be described by the standard deviation (s.d.). For a single person, his or her weight will be 70 ± 13 kg, ignoring gender differences. Here the ± value is the standard deviation calculated from the statistics of people's weight. Moreover, as the number of people increases, the standard deviation of the mean tends to decrease. In fact it can be shown that from the central limit theorem of statistics[18] that the standard deviation of the weight per person decreases in a predictable way as the number of people increase. For our aeroplane, the mean weight per person is $70 \pm 13/\sqrt{n}$ kg for n people. For modern jet airliners, $n \approx 250$, so $\sqrt{n} \approx 15$. The uncertainty in the total weight of people is only $13\sqrt{n} \approx 200$ kg. However, the weight of the baggage might vary enormously from flight to flight. So the airlines weigh the baggage and estimate the weight of the people. Hence, the total load on the aircraft for 250 people carrying 20 kg of baggage each, is on average (20 + 70) x 250 = 22,500 kg, and this figure will have an error of about 200 kg (error < 1%), well within operating tolerances.

7.8 Allometric Scaling in Biology

The BMI is one of a number of specific parameters that can be applied to humans and wildlife. We have already see in § 2.8 how simple scaling can be used to estimate the mass of a mouse when we know only its length. All that we were doing was effectively applying a specific density to an object of a specific shape and since we were only concerned about generalities, any general mammal shape could be used as a starting point. However, this procedure isn't entirely accurate. Essentially, when an animal becomes larger its bones need to support more weight and so the weight itself increases with length, so the strict scaling in § 2.8 doesn't apply. For example, to take the BMI case, using scale analysis directly would give an index (representative of density) ~ Mh^{-3} for a person's mass M and height h. We find that taking account of body shape empirically leads to the BMI = Mh^{-2}, so the scaling is less marked on height than would be expected. Finding relationships of this kind, and relationships between species is the discipline of allometry[19].

Interspecies investigations have led to the discovery of Kleiber's law [20] in which an animal's metabolic rate scales approximately as the ¾ power of the mass of the animal. Application of sound physics to biology also leads to conclusions that giant insects and spiders, larger than a person, beloved of horror films, are impossible. The reason is that the larger an object the smaller is the ratio of the surface area to volume. An insect 'breathes' in through spiracles which are basically holes in the skin and the oxygen comes in the holes. As the insect gets larger so this becomes inefficient as the organs in the bulk of the insect cannot be reached[21]. So there is a natural limit to the size of insects and spiders of order 10 cm, not the several metres you sometimes see in films. Another meaningless comparison that you see in the media are the 'remarkable' feats of strength performed by ants and fleas. A flea is quite capable of jumping 15 cm up in the air[22, 23] because the energy expended on take off is roughly proportional to the flea's mass and the potential energy is also proportional to mass. It doesn't make particular sense to compare the 50-100 times body length performance of a flea (the largest of any species) with the "puny" 1.25 body length performance of a professional high jumper. An ant can lift relatively heavy weights since the strength to weight ratio is inversely proportional to size[24], giving a relative advantage to small animals, and insects in particular. A 2t load on an adult human would considerably stress the joints. By proper investigation of the physics of load bearing the proper relationship between physical performance and animal size can be explored, rather than taking the face-value idea that some insects are somehow "super strong".

7.9 Density and Specific Gravity

Density is the mass per unit volume and can be used to measure any material. Specific gravity is the density relative to water, and is used primarily for liquids for which small deviations from 1.0 (pure water) are important. For example, the specific gravity of beer and wine are measured to estimate the alcohol content and this is often carried out using a hydrometer[25]. A hydrometer is a tube-shaped instrument which is calibrated to float at a level indicating the density of the liquid. It essentially uses the principle discovered by Archimedes several millennia ago that the mass of liquid displaced is equal to the mass of a floating object[26].

Densities vary through many orders of magnitude, depending primarily on whether the material is gas, liquid or solid. Densities are also weakly temperature dependent and tend to reduce as the temperature increases. This is because at higher temperatures, molecules vibrate more and their separation distance increases in the average. Water is an exception in that its density is a maximum at 4 °C, and decreases between 4 °C and the freezing point. Starting with a solid of any general material, and increasing the temperature, the density almost invariably decreases, undergoing a step change at the melting point. As a liquid, the density continues to decrease with increasing temperature until the boiling point is reached. The density then decreases substantially as the liquid evaporates to a gas. As a gas, the material follows very nearly the ideal gas law, with the density inversely proportional to the temperature measured in kelvins (Chapter 8). The change of the density of a liquid with temperature is often exploited in the design of thermometers (Chapter 8). Solids respond much less to temperature change than liquids or gases, but this expansion is also exploited in the design of bi-metallic strips for temperature control. (Chapter 8).

Measuring the World, by John Austin

Density units in fps are lb ft^{-3} (pounds per cubic foot). The metric (SI) unit is kg m^{-3} (kilogrammes per cubic metre). In the cgs system, densities are in g cm^{-3}. Many people may have been brought up on those units, as I was originally, but it is easy to switch to kg m^{-3}. Values are a bit unfamiliar at first (1000 kg m^{-3} for water), but you get used to the new numbers quickly. 1 kg m^{-3} = 2.205 x 0.3048^3 = 0.06243 lb ft^{-3}. 1 lb ft^{-3} = 16.02 kg m^{-3}. In other words, an approximate conversion from kg m^{-3} to lb ft^{-3} is to multiply by 15 (error 7%) or better still 16 (error 0.1%). To convert from lb ft^{-3} to kg m^{-3} multiply by 6 and divide by 100 (error 3%).

Densities of substances vary from below 0.1 to over 20,000 kg m^{-3} at standard atmospheric pressure. The lowest density substance is hydrogen gas, which at standard temperature and pressure (STP; 0 °C, 1 atmosphere pressure) has a density of 0.09 kg m^{-3}. Air at STP has a density 1.3 kg m^{-3}. The densest gas containing naturally occurring elements is uranium hexafluoride, which we met in § 4.6. It has a density at STP of 15.5 kg m^{-3}. As noted in § 4.6, the gas is used in centrifuges to separate uranium isotopes for bomb construction and nuclear power.

At STP or near room temperature, typical liquid densities are about the same as water, 1000 kg m^{-3}[27]. Alcohol has a density of 785 kg m^{-3}. The densities of light oils and gasoline are about 800 kg m^{-3}, while crude oil has a density of between 790 and 970 kg m^{-3}. The heaviest density liquid is mercury (13,534 kg m^{-3} at 20 °C). It is also unique in being the only liquid metal element at room temperature while gallium, a semi-conductor, melts at 30 °C, somewhat above room temperature. The non-metallic element bromine is a also a liquid at room temperature (melting point -7 °C), but is highly volatile, with a boiling point of 59 °C[28].

Solid densities vary considerably. Some solids, such as

polystyrene, have gases pumped into their structure, so their densities can be very low. Certainly, expanded polystyrene is manufactured with densities in the range 10-50 kg m^{-3}[29], 1-5% of the density of water. Of the elements, lithium has a low density (534 kg m^{-3})[30], less than water. However, it is extremely reactive and doesn't occur in nature in its elemental form. The densest naturally-occurring element is osmium (22,590 kg m^{-3}) although artificially produced elements have still higher densities.

References

[1] The BIPM watt balance → towards a definition of the kilogram, Bureau International des Poids et Mesures, Sèvres, France, http://www.bipm.org/en/scientific/elec/watt_balance/, accessed 10 March 2014.
[2] Water, Wikipedia, 10 March 2014, http://en.wikipedia.org/wiki/Water, accessed 10 March 2014.
[3] Water, How stuff works, http://science.howstuffworks.com/dictionary/geology-terms/water-info.htm, accessed 10 March 2014.
[4] Water management in the steel industry: report now available, Press Release, Worldsteel Association, 15 May 2011, http://www.worldsteel.org/media-centre/press-releases/2011/water-management-report.html, accessed 10 March 2014.
[5] Kilogram, Wikipedia, http://en.wikipedia.org/wiki/Kilogram, accessed 10 March 2014.
[6] Ancient Roman units of measurement, Wikipedia, 27 January 2014, http://en.wikipedia.org/wiki/Ancient_Roman_units_of_measurement, accessed 10 March 2014.
[7] Avoirdupois, Wikipedia, 3February 2014, http://en.wikipedia.org/wiki/Avoirdupois, accessed 11 March 2014.
[8] Pound (mass), Wikipedia, 11 February 2014, https://en.wikipedia.org/wiki/Pound_(mass), accessed 11 March 2014.
[9] Bureau International des Poids et Mesures, Pavillon de Breteuil, F-92312, Sèvres, France, http://www.bipm.org/, accessed 11 March 2014.
[10] Gravitation of the moon, Wikipedia, 23 January 2014,

http://en.wikipedia.org/wiki/Gravitation_of_the_Moon, accessed 11 March 2014.
[11] Slug (unit), Wikipedia, 24 January 2014, http://en.wikipedia.org/wiki/Slug_(mass), accessed 11 March 2014.
[12] Carat (unit), Wikipedia, 21 February 2014, http://en.wikipedia.org/wiki/Carat_(mass), accessed 11 March 2014.
[13] Carat (purity), Wikipedia, 6 March 2014, http://en.wikipedia.org/wiki/Carat_(purity), accessed 11 March 2014.
[14] Analytical balance, 26 June 2009, author: Sarcyn, http://commons.wikimedia.org/wiki/File:Analyticalbalance1.jpg, accessed 11 March 2014.
[15] Weighing scale, Wikipedia, 7 March 2014, http://en.wikipedia.org/wiki/Weighing_scale, 11 March 2014.
[16] BMI healthy weight calulator, NHS Choices, 31 October 2013, http://www.nhs.uk/Tools/Pages/Healthyweightcalculator.aspx, accessed 11 March 2014.
[17] Mass of all the people on Earth, Daniel Touger, 2006, The Physics fact book, Hypertextbook, edited by Glenn Elert, http://hypertextbook.com/facts/2006/DanielTouger.shtml, accessed 12 March 2014.
[18] Central limit theorem, Wikipedia, 14 February 2014, http://en.wikipedia.org/wiki/Central_limit_theorem, accessed 12 March 2014.
[19] Allometry, wikipedia, 1 February 2014, http://en.wikipedia.org/wiki/Allometry, accessed 12 March 2014.
[20] Kleiber's law, Wikipedia, 25 February 2014, http://en.wikipedia.org/wiki/Kleiber's_law, 12 March 2014.
[21] What keeps bugs from being bigger, Argonne National Laboratory, 8 August 2007, http://www.aps.anl.gov/Science/Highlights/2007/20070808.htm, accessed 12 March 2014.
[22] Cadierques, M.C. et al. 2000: A comparison performance of the dog flea, Ctenophalides canis (Curtis, 1826) and the cat flea, Ctenophalides felis felis (Bouché, 1835), Vet Parasitol, 2000, Oct 1, 92(3), 239-241.
[23] Flea, Wikipedia, 9 March 2014, http://en.wikipedia.org/wiki/Flea, accessed 12 March 2014.
[24] How can ants lift objects 50 times their body weight?, Debbie Hadley,

About.com insects, http://insects.about.com/od/antsbeeswasps/f/ants-lift-50-times-weight.htm, accessed 12 March 2014.

[25] Hydrometer, 15 February 2014, Wikipedia, http://en.wikipedia.org/wiki/Hydrometer, accessed 12 March 2014.

[26] Archimedes principle, Wikipedia, 6 March 2014, http://en.wikipedia.org/wiki/Archimedes'_principle, accessed 12 March 2014.

[27] Liquids – densities, The engineering toolbox, http://www.engineeringtoolbox.com/liquids-densities-d_743.html, accessed 12 March 2014.

[28] Bromine, Wikipedia, 18 February 2014, http://en.wikipedia.org/wiki/Bromine, accessed 12 March 2014.

[29] EPS Technical information, http://www.isoclad.co.uk/pdf/EPS_Datasheet.pdf, accessed 12 March 2014.

[30] Densities of the elements (data page), Wikipedia, 22 January 2014, http://en.wikipedia.org/wiki/Densities_of_the_elements_(data_page), accessed 12 March 2014.

Measuring the World, by John Austin

Measuring the World, by John Austin

8. Temperature

Temperature = The degree of hotness or coldness of a body: heat transfers from a hot to a cold body

8.1 Introduction

The importance of temperature as a physical quantity was recognised long before its fundamental nature was fully understood. Many processes in nature occur faster at higher temperatures and slower at lower temperatures. Temperature also controls many observed features of nature. For example, water freezes at low temperatures and boils at high temperatures. We have seen in Chapter 7 how temperature influences the density of materials. The expansion of liquid with temperature, in particular, has been used to establish a temperature scale, and by the 18th century, the traditional instrument to measure temperature was the mercury-in-glass thermometer[1].

By the 19th century the fundamental nature of temperature was recognised as a measure of the amount of vibration of the atoms in the structure of solids, and a measure of the kinetic energy of atoms in gases and liquids. For gases, this was a powerful concept which supported the kinetic theory of gases, a theory that was confirmed experimentally[2]. This established temperature as a rigorous measurement which enabled the SI to declare the preferred unit of kelvins (K), related to Celsius, °C (§ 8.4). In practice, temperatures vary between the extremely low values of outer space (2.7 K)[3] to the very high values in the core of stars such as the sun, estimated to be 15 million K[4]. Almost the whole range of temperatures is attainable in the physics laboratory, even if only for a fleeting time. Temperatures closer to absolute zero have been attained, with a world record now of just 100 pK (10^{-10} K), achieved in 1999[5]. Current maximum laboratory temperatures now exceed 3.7 GK[6], which was achieved by the Sandia Z machine used for nuclear fusion research. Consequently, we can explore the properties of matter, as they change substantially with temperature. Life as we know it, can only survive over a very limited range of ambient temperatures, which has an important bearing on the existence of life elsewhere in the universe. This is variously reported as the Goldilocks effect or principle after the young girl in the nursery rhyme who finds one of the bowls of porridge just at the right temperature to eat[7].

In § 8.2, the notion of a temperature scale is introduced. In § 8.3 the properties of gases are described and how these led to a fundamentally new temperature scale, the thermodynamic temperature scale. Conversions between temperature scales are given in § 8.4, and the thermodynamic temperature scale is explained in further detail in § 8.5. The design of thermometers is explored in § 8.6. Atmospheric temperature and measurements are discussed in § 8.7 – 8.9, finishing with a discussion of the greenhouse effect and climate change.

8.2 Celsius and Fahrenheit

To establish a temperature scale requires the measurement of some physical property which is taken to vary linearly with temperature between two fixed points. For example, if the volume of a liquid is V_1 at temperature T_1, and V_2 at a higher temperature T_2, then at some intermediate volume V, the temperature, T, is given by the following expression.

Measuring the World, by John Austin

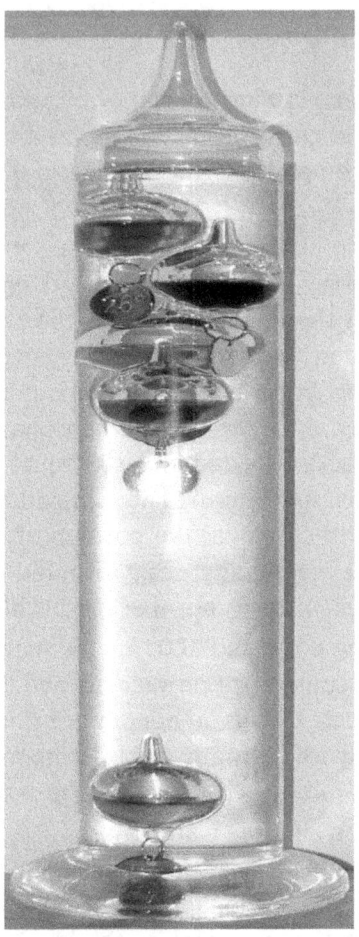

Galilei thermometer, 3 January 2005, author: Grin. From Wikimedia[8]. This amazingly beautiful thermometer is a modern and functional version of one named after, but not designed by Galileo Galilei[9]. As the temperature changes, the clear fluid in the tube changes in density and the coloured globules float or sink to indicate the temperature. Such thermometers can be bought nowadays for a modest expenditure in stores such as Tescos in the UK.

Measuring the World, by John Austin

$$T = T_1 + (T_2 - T_1)\frac{(V - V_1)}{(V_2 - V_1)}$$

To simplify matters, suppose that we have an uncalibrated standard mercury-in-glass thermometer. The thin capillary tube of the thermometer emphasises the change in volume at the base (the bulb). We can choose a fixed point of 0° for the melting point of ice. So, we would place our thermometer into a flask of ice just at the point of melting. We then make a mark on our thermometer where the edge of the liquid appears in the capillary tube. Then repeat the operation with the thermometer in a pan of water which is just boiling. We then draw a mark at the position of the new meniscus level. Once the thermometer has cooled down, we can mark a linear scale, divided into 100 equal 'degrees' between the two fixed points. Any intermediate temperature would be read from the scale. With care over the position of our 2 fixed points, we would have the 'centigrade' scale invented in 1742 by Anders Celsius[10]. However, it turns out that Celsius himself gave a value of 0° for the boiling point and 100° for the melting point. In 1948 the Ninth General conference on Weights and Measures renamed the temperature scale Celsius, in honour of its inventor[11]. Rather than using melting ice, a more stable temperature is the triple point of water, occurring at 0.01 °C[12]. The triple point refers to the conditions when all three water phases (ice, water, vapour) are in equilibrium and this occurs at 0.00604 Atmospheres pressure. There is a further complication in that the boiling water fixed point is dependent on ambient pressure, so this is set at 1 Atmosphere.

The Fahrenheit scale was introduced in 1724[13], but is less well documented. For this scale, Fahrenheit, a German physicist, chose a lower fixed point (0 °F) as the lowest attainable freezing point for a salt and water mixture. Unfortunately, the exact amounts of salt and water were not recorded. Documentation on the upper fixed point appears to be missing, so many of the accounts of the Fahrenheit scale could be misleading.

Measuring the World, by John Austin

Most accounts suggest that the upper fixed point was taken to be 212 for the boiling point of water but this seems unlikely, and was likely added after the centigrade scale was in common use. A more likely possibility is that the upper fixed point was 100 °F for body temperature, as suggested by Wikipedia[13], but we now know that this figure is slightly too high. In any case, body temperature varies naturally throughout the day by as much as 1 °F in a healthy individual. Now °F is entirely defined in terms of °C, Celsius, and it is in my view time for countries like the USA to abandon it.

8.3 The Properties of Gases

During the 18th and 19th centuries, laboratory work with gases made substantial progress in understanding temperature, building on earlier work by Boyle. The scientists Charles, Guy-Lussac and Avogadro showed by a sequence of experiments that for a fixed mass of gas, the equation[14] relating its properties is

$$PV = mR(T + \alpha)$$

P is the pressure, V is the volume, m is the mass of gas and T its temperature (°C). R is a constant which varies according to the gas (see Chapter 13). The equation is deceptively simple in today's notation, but when the work was being done, coordinating the set of experiments into a single equation would have been quite challenging bearing in mind the disparate set of units then in existence. In modern terminology, it was found that $\alpha = 273$ °C for all gases. Hence the equation predicted that the pressure or volume of a fixed mass of gas would go to zero at a temperature T = - 273 °C. This is a very special temperature known as 'absolute zero'. Many exotic properties of materials occur at very low temperatures (§ 8.5). Making the assumption that T + 273 °C is a measure of the kinetic energy of gas molecules, the kinetic theory of gases has been able to reproduce the properties of gases, including the gas equation using idealised assumptions. Furthermore, *T* + 273 °C appears in a wide range of physical processes and is referred to as the thermodynamic temperature. It is extensively used in scientific work.

Thermodynamic temperature has units of kelvins (K). It is written without the degree sign to indicate that, like the measurement of length or mass, it is an absolute measurement. The scale has two fixed points: 0 K (- 273.15 °C), when all motion ceases, and the triple point of water at 273.16 K (0.01 °C). The property used to define temperature is the behaviour of an ideal gas, one which obeys $PV = mRT$ exactly where T is now in kelvins. To measure the temperature in practice, corrections may need to be made to the physical behaviour of real materials (§ 8.5). The Fahrenheit scale no longer has any independence and is defined in terms of the Celsius scale (0 °C = 32 °F, 100 °C = 212 °F).

8.4 Temperature Conversions

The simplest way to write this is:

°F = 32 + 1.8 °C   °C = (°F - 32)x5/9   K = °C + 273.15

Because of the large addition to go from °C to K, °C is often used even in scientific work. Temperature intervals in °C are the same as temperature intervals in K, so intervals are more properly written in K. For example, the increase in temperature from melting ice to boiling water is 100 °C or 100 K. As noted in § 8.3, the degree sign for K is expressly forbidden, but that doesn't stop some scientists, who should know better!

Throughout my life, I have been regularly exposed to measurements of temperature is °F and °C, especially for the weather. As a result, faced with a °F temperature, I can readily convert to °C within a few seconds using my memory of the exact conversions 23 °F (-5 °C), 32 °F (0 °C), 41 °F (5 °C), 50 °F (10 °C), 59 °F (15 °C), 68 °F (20 °C), 77 °F (25 °C), 86 °F (30 °C), and 95 °F (35 °C). Outside that range, I have to think a bit harder. Note that -40 °F is the same as -40 °C which can be exploited in the Arctic or Antarctic if you venture that far! Far better than remembering conversions is to remember for example what 15 °C feels like on your skin and then to forget Fahrenheit entirely. While I can do that myself, society rarely allows me to.

The responses of the public about temperature are amusing, unfortunately in Britain often sensationalised by the press. If you ask an American for a temperature estimate, without specifying the units, the answer will come in °F, even from an American scientist! The British used to be confused and may well still be. On a warm day, the answer might be given in °F (e.g. 80°) [15]. But even the Daily Mail quotes high temperatures in Celsius these days[16]. There's something sort of endearing about the British weather if 27 °C is considered headline news! On a very cold day, the answer from the public will almost certainly be given in °C (e.g. -5 °C). In the latter case, 0 °C is a significantly emotive turning point, so once the temperature drops below 10 °C, the allure of Celsius becomes stronger!

8.5 The Thermodynamic Temperature Scale

In the kinetic theory of gases[2], the mean kinetic energy of each molecule is $½mv^2 = 3kT/2$. The mass of each molecule is m, and v is the mean molecular velocity in the gas; k is Boltzmann's constant (Chapter 2) and T is temperature. The factor 3 on the right hand side of the equation arises from the inclusion of energy from the velocity in each of the 3 spatial dimensions. With the above equation, the kinetic theory of gases proceeds to deduce the ideal gas law[2], providing T is in kelvins.

Therefore, the SI system uses the ideal gas law to define temperature. In practice, real gases diverge slightly (about 1%) from the gas law, but the specification of temperature fixed points ensures reproducibility of the scale from 0 to 273.16 K. Above this temperature, secondary standards are used to provide additional fixed points. These fixed points have been defined by the International Temperature Scale of 1990[17] and are primarily for industrial use. Unlike in the centigrade scale, the boiling point of water is not used as a fixed point, as the boiling temperature is too dependent on pressure to provide a reliable fixed point. Instead, the temperatures include the melting point of gallium (302.9146 K, 29.7646 °C) and the freezing points of the metals indium, tin, zinc, aluminium, silver gold and copper (1357.77 K, 1084.62 °C). All of these fixed points may be subject to adjustment if the temperature scale becomes better known, or when changes to the SI take place (Chapter 17).

Note that a mercury-in-glass or alcohol-in-glass thermometer will deviate slightly from the ideal gas temperature, which is the SI primary standard. In other words, the temperature dependence of mercury expansion, or alcohol expansion is not quite linear with thermodynamic temperature. This is usually an extremely small effect, but for high temperatures the error can approach several K when using secondary standards. For this reason, as well as specifying the secondary fixed points, the ITS-90[17] also specifies the type of thermometer to be used and exactly how the physical property is to be interpolated between fixed points. See also the figure (below). The above expression indicates that molecular energy is zero at absolute zero ($T = 0$ K). Strictly, whatever sample of material we have will always have some energy, however small. Actually reaching 0 K is therefore impossible but we can get very close. Laboratory temperatures, for example, have reached 1.0×10^{-10} K [8]. At low temperatures, exotic physical properties arise, including flow without friction in a

Measuring the World, by John Austin

'Bose-Einstein Condensate'[19], and superconductivity[20]. A most bizarre property of superfluid helium at low temperatures is that it can flow through 'solid walls'[21], namely its 'solid' container.

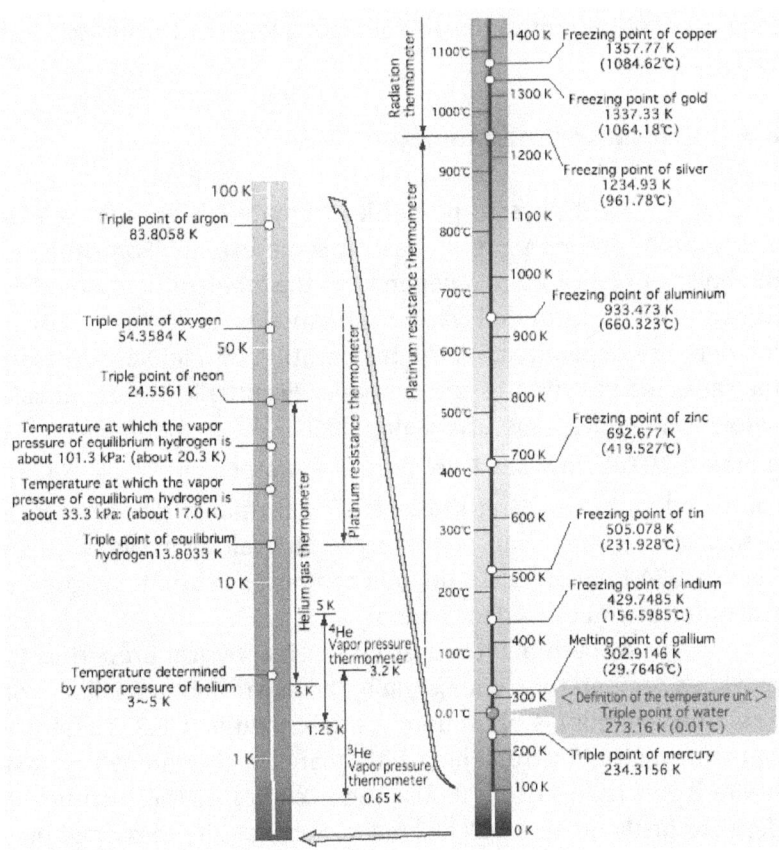

The thermodynamic temperature scale, indicating the fixed points agreed by ITS-90. Figure from the National Metrology Institute of Japan[18].

Temperature is also a measure of the equilibrium levels of electrons around atoms. If there is an energy inversion, with electrons appearing in higher energy levels without filling the lower energies first, then the temperature of the atom (in K) could be defined as negative[22]. Nonetheless, this is usually a temporary state prior to the discharge of a laser, or the emission of radiation.

8.6 The Design of Thermometers

Most physical properties of materials are affected by temperature and pressure at least to some extent. For example, the boiling point of a liquid depends on the ambient pressure. This is because the saturated vapour pressure (svp) (Chapter 10) is temperature dependent. As the temperature of a liquid increases, the saturated vapour pressure increases. Eventually, the svp equals ambient pressure, and the liquid boils. Up a mountain, at a pressure 20% below sea level pressure, water boils at 94 °C[23], compared with 100 °C at sea level. Trivial though this difference seems, if you are living or visiting a high altitude city such as Denver, USA (altitude 1615 m), you can never get quite as good a cup of tea as you can get in London!

A thermometer uses one of its physical properties to provide a temperature. For example, the expansion of a volume of alcohol or mercury is often used, as described in § 8.2. These are commonly used instruments, with mercury thermometers first having been used in the 1600s. The accuracy of the instrument depends firstly on how well it is made. Secondly, it is assumed that the expansion occurs linearly with temperature (§ 8.2), and there may be a small divergence between the mercury-in-glass temperature and the ideal gas temperature. Clearly linear relations do not occur over all temperatures. For example, trivially, above the boiling point of alcohol, an alcohol thermometer would have become unreliable, with gaseous and liquid alcohol filling the thermometer. Other thermometers use electrical resistance as a physical property, e.g. the platinum resistance thermometer[24]. Thus a variety of temperatures are possible for the same conditions, and the resulting measurement would be known as the 'alcohol-in-glass', the 'mercury-in-glass' or 'platinum resistance' temperature. In very precise work, corrections may need to be made to convert to 'ideal gas temperature' which is the SI primary standard. Certainly at temperatures comfortable to living creatures, 0 – 40 °C, only a small change (15%) in thermodynamic temperature is implied, and corrections are of the order of 0.001 K[25], which can be ignored for many purposes. However, at high temperatures, 850 °C, a property such as platinum resistance may need a correction of about 0.4 K[25] and radiation thermometers[26] need to be used.

Pyrometers[27], or radiation thermometers, measure either the total radiation from an object, or the wavelength at which the radiation peaks. The colour of a glowing object is a measure of this peak wavelength. The theory of radiation transfer was established by Planck and others, and the Planck function is the functional form of the radiation emitted at a specific wavelength for a body at a given temperature[28]. The theory was established at the beginning of the 20th century and was one of the triumphs of the new field of quantum mechanics. It is somewhat complicated, but details can be read here[28]. An incandescent light bulb with a filament emits radiation peaking in the orange part of the spectrum, showing that the filament temperature is about 3000 K. The surface of the sun emits radiation peaking near yellow at 585 nm wavelength, so that the surface temperature is about 6000 K. All objects emit radiation in the electromagnetic spectrum, with the amount increasing with the temperature of the objects. The total energy radiated per unit area over all wavelengths is σT^4, assuming that the material surface is not reflective. The value of σ is derivable from the Planck theory and is known as the Stefan-Boltzmann constant. It is given in Chapter 2 (5.67037×10^{-8} W m^{-2} K^{-4}). The peak wavelength is inversely proportional to temperature, and is known as the Wien displacement law[28]. For a body like the Earth, which has a mean surface temperature of 288 K (15 °C), it emits mainly in the infrared. This is absorbed by CO_2 in the lower atmosphere. As the CO_2 increases, so does atmospheric temperature, because of the extra absorption. This is the origin of the 'greenhouse effect', although confusingly the processes operating in a greenhouse are different, but the name has stuck.

Other types of thermometers are in common use. Bimetallic thermometers[29] use two different metals with different thermal expansion rates. They are often used in the shape of a coil, and as the temperature changes, so does the angle of a pointer at the end of the coil. The pointer indicates the temperature change, or triggers an electrical circuit, in the case of a thermostat. In a thermocouple[30], two different wires are joined together. When a temperature difference is applied to the two ends, an electrical current flows. The temperature difference is determined from the current flow. This instrument is designed to give temperature difference but by proper calibration can give absolute temperatures. Liquid crystal thermometers[31] are also available. The crystals are often attached to a card, and as the temperature changes, so does the colour of the crystals. As well as platinum, resistance thermometers have been made from various metals, such as nickel and copper[24], or semi-conductors such as silicon.

8.7 Atmospheric Temperature

The study of atmospheric temperature covers a very broad area which is well documented[32]. Essentially, heating by

Measuring the World, by John Austin

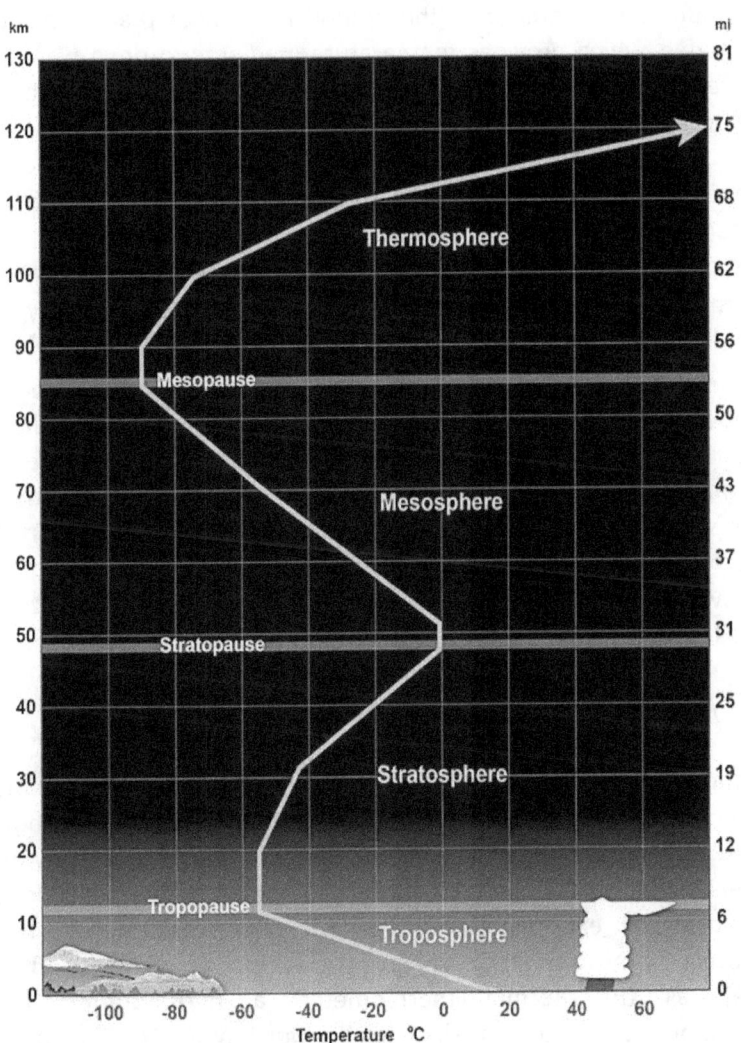

The change in temperature of the atmosphere with height from the US National Weather Service[33]. The temperature variation with height divides the atmosphere into distinct layers.

the sun at the Equator at the ground is stronger than at higher latitudes, which provides temperature gradients to drive weather systems. The heat absorbed by the oceans is redistributed globally by ocean circulations.

Since the sun's radiation is absorbed at the ground, atmospheric temperature falls with height, at a typical rate of 6.5 K for every kilometre increase in height. In contrast, the upper atmosphere contains ozone (O_3) formed from the breaking up of normal oxygen (O_2) by solar ultra-violet (UV) to form oxygen atoms. Each O atom then reacts with an O_2 molecule to form O_3. The process is exothermic, and heats the upper atmosphere. Above about 10 km altitude in high latitudes, but at an altitude of nearly 16 km above the tropics, the temperature starts to increase with height. At an altitude of about 50 km, ozone concentrations are small, and the heating effect is small, so temperatures start to decrease again with height. The region up to 50 km altitude is known as the stratosphere, where high concentrations of ozone are found.

8.8 Atmospheric Measurements

Atmospheric measurements are made with surface instruments, balloons, aircraft and satellites[34] and here we discuss only those that refer to temperature near the surface. Thermometers were mentioned in § 8.6. For accurate measurements, the instruments need to be exposed within properly designed screens which protect them from direct sunlight, and excessively strong wind. As well as the normal thermometer, a maximum-minimum thermometer is included. The analogue version of this[35] contains two thermometers with their capillary ends joined in a U-bend. In the bend is contained a column of mercury and the mercury drags with it small metal pins, one at each end of the column. The thermometer uses alcohol as the fluid, and as it expands, it pushes the mercury column which drags the pin with it. At the highest temperature, the pin is left in place with the alcohol column shrinking to the new, lower, temperature. The minimum temperature works on the same principle, with the second metal pin. The instrument is left for the period of interest, say 24 hours, and at the end of the time, the instrument is examined. The positions of the pins are recorded, indicating the maximum and minimum temperature, and the pins are reset using a magnet. Nowadays, digital devices[35] do the job in a way that is much more satisfactory and allow the user to download the data to his or her PC for analysis.

Humidity is determined by placing a cloth, dipped in water, over the bulb of one thermometer. The temperature recorded is the wet bulb temperature and the humidity can be derived by comparing it with the dry bulb temperature. In dry air, water evaporates from the cloth or wick, and cools the thermometer. The larger difference in temperatures, the lower the humidity. When the temperatures are the same, the humidity is 100%. Digital meters also offer improved methods of measuring humidity[35, 36].

Generally, we find a combination of conditions (temperature, humidity, wind) affects our degree of comfort. This is especially true in continental climates such as experienced by the USA or Europe. Accordingly, many National Weather Services issue warnings of heat index and wind chill. In the summer, high temperatures combined with high humidity can cause severe discomfort. In the winter, low temperatures combined with strong winds can cause discomfort at the opposite end of the spectrum.

8.9 The Greenhouse Effect and Climate Change

The Earth's atmosphere provides a natural greenhouse effect without which we wouldn't survive but there is also a man made greenhouse effect which can be readily estimated. However, If we were suddenly to stop producing greenhouse gases, the climate would take many decades to come into balance. Because of the increasing greenhouse gases present in the atmosphere, the energy from the sun exceeds the energy emitted by the Earth as heat by a small amount, about 1 W for every square metre of the Earth's surface. This energy is passed to the deep oceans and over decades the sea temperature will rise. If by a miraculous occurrence all greenhouse gas emissions into the atmosphere were suddenly stopped, the warming of the atmosphere and ocean would continue until the heat flux into the oceans became balanced and the net heat emitted to space also went to zero, thereby balancing the climate. Calculations suggest that the additional warming would be about 0.5 °C [37].

If the Earth had a balanced climate, the heat received from the sun would be balanced by heat emitted. According to the physics of black body radiation[28], this gives $E = (1-e)\sigma T^4$, where e is Earth's albedo (reflectivity, 0.3), σ is the Stefan-Boltzmann constant (5.67×10^{-8}) and T is the global average temperature. The energy from the sun, E, can be measured (341 W m^{-2}), and with the known value of σ, we arrive at a temperature, $T = 255$ K $= -18$ °C. This compares with the average temperature of the Earth of 15 °C. So, the atmosphere, and the greenhouse effect are responsible for 33 °C of additional warming. Most of this is due to water vapour, with a smaller but important effect due to carbon dioxide, CO_2 and other greenhouse gases. The impact of GHGs in the atmosphere are calculated to be as follows[38]:

H_2O (36-72%), CO_2 (9-26%), CH_4 (4-9%), O_3 (3-7%)

In the pre-industrial age, the CO_2 amount was natural and the energy input was exactly balanced by the cooling of the Earth to space. Once more CO_2 was added into the atmosphere due to human activities, an imbalance of about 1 W m^{-2} of the Earth's surface has developed. This is rather small, at only 0.3% of the input energy from the sun, but it is crucial. A certain amount of further global warming is unavoidable because of the 0.3% energy imbalance[37]. Global change will finally stop at a temperature which depends mainly on the amount of CO_2 released into the atmosphere since industrialization[39].

As noted above, of the 33 K greenhouse effect, about 10%, or 3 K is due to naturally occurring CO_2. However, humans have already contributed a further 48% CO2, increasing the concentration from 270 pre-industrially to about 400 parts per million (ppm)[40]. It follows that a substantial amount of the CO_2 greenhouse effect i.e. about 1 K, has been human induced as noted by in depth studies[41]. Further inevitable CO_2 increases will raise global temperature by at least another 1 K and perhaps by 2 K or more. However, the effects will vary around the world. For example, the poles are warming much faster than the rest of the surface. Unfortunately, for some prominent US politicians, scientists don't seriously question the likely future climate change, or that humans are responsible. Rather, current research is aimed at clarifying local changes, e.g., the extent of future polar warming. There are suggestions for example, the Arctic ice cap may be melting at a faster rate than suggested by most computer predictions.

The Stern report[42] indicated that climate change harms the economy. Other analyses seem less clear cut[43] but a full analysis should include both the harmful effects of climate change and the problems of a carbon-based economy, e.g., air pollution. At higher temperatures, people work less well and are less productive, points not concerned by the Stern report. Consequently, climate change is probably already a cost to society even though it wasn't expected to become a net cost in decades. There is a distinct possibility that climate change costs more than the emission cuts needed to tackle them. Caps on emissions, for example, could be tighter, and carbon prices could be much higher than they are at present. Unfortunately, greenhouse gases are a global problem, because they become uniformly mixed in the atmosphere, but global solutions require local action. Many countries such as Canada, China and the USA seem to have decided that it is not their problem.

References

[1] History of temperature, CAPGO, http://www.capgo.com/Resources/InterestStories/TempHistory/TempHistory.html, accessed 12 March 2014.
[2] Kinetic theory, Wikipedia, 9 March 2014, http://en.wikipedia.org/wiki/Kinetic_theory, accessed 12 March 2014.
[3] Cosmic background explorer, D.T. Chuss, 26 June 2008, http://lambda.gsfc.nasa.gov/product/cobe/, accessed 12 March 2014.
[4] Sun fact sheet, D.R. Williams, NSSDC, NASA Goddard space Flight Center, Greenbelt, MD, USA, 1 July 2013, http://nssdc.gsfc.nasa.gov/planetary/factsheet/sunfact.html, accessed 12 March 2014.
[5] World record in low temperatures, T. Knuuttila, O.V. Lounasmaa Laboratory, 8 December 2000, http://ltl.tkk.fi/wiki/LTL/World_record_in_low_temperatures, accessed 12 March 2014.
[6] Z machine, Wikipedia, 7 March 2014, http://en.wikipedia.org/wiki/Z_machine, accessed 12March 2014.
[7] Goldilocks principle, Wikipedia, 3 February 2014, http://en.wikipedia.org/wiki/Goldilocks_principle, accessed 12 March 2014.
[8] File: Galilei homero.jpg, Wikimedia, 3 January 2005, http://commons.wikimedia.org/wiki/File:Galilei_homero.jpg, accessed 12 March 2014.
[9] Galileo thermometer, Wikipedia, 2 March 2014, http://en.wikipedia.org/wiki/Galileo_thermometer, accessed 12 March 2014.
[10] Celsius, Wikipedia, 11 March 2014,

Measuring the World, by John Austin

http://en.wikipedia.org/wiki/Celsius, accessed 12 March 2014.
[11] CIPM, 1948 and 9th CGPM, 1948, Adoption of "degree Celsius", Bureau International des Poids at Mesures, Sèvres, France, http://www.bipm.org/en/committees/cipm/cipm-1948.html, accessed 12 March 2014.
[12] Triple point, Wikipedia, http://en.wikipedia.org/wiki/Triple_point, accessed 12 March 2014.
[13] Fahrenheit, Wikipedia, 12 March 2014, http://en.wikipedia.org/wiki/Fahrenheit, accessed 13 March 2014.
[14] Gas laws, Wikipedia, 6 March 2014, http://en.wikipedia.org/wiki/Gas_laws, accessed 13 March 2014.
[15] A very British mess, A report by the UK Metric Association, 2004, Chapter 7, http://www.metric.org.uk/avbm-summary, accessed 13 March 2014.
[16] Summer's here! Today is the hottest day of the year so far with temperatures reaching 27C... but tomorrow will be EVEN HOTTER, Daily James Tozer, Daily Mail, 4 July 2013, http://www.dailymail.co.uk/news/article-2356392/UK-weather-Today-hottest-day-year-far-temperatures-reaching-27C—tomorrow-EVEN-HOTTER.html, accessed 13 March 2014.
[17] International temperature scale of 1990, Wikipedia, 12 December 2013, http://en.wikipedia.org/wiki/International_Temperature_Scale_of_1990, accessed 13 March 2014.
[18] Standards for the SI base units, The definition of the kelvin and the International temperature Scale, https://www.nmij.jp/english/library/units/temperature/, accessed 13 March 2014.
[19] Bose-Einstein condensate, Wikipedia, 11March 2014, http://en.wikipedia.org/wiki/Bose%E2%80%93Einstein_condensate, accessed 13 March 2014.
[20] Superconductivity, Wikipedia, 8 March 2014, http://en.wikipedia.org/wiki/Superconductivity, accessed 13 March 2014.
[21] Superfluid helium, Youtube, http://www.youtube.com/watch?v=2Z6UJbwxBZI, accessed 13 March 2014.
[22] Negative temperature, Wikipedia, 7 February 2014, http://en.wikipedia.org/wiki/Negative_temperature, accessed 13 March

2014.
[23] Water – pressure and boiling point, http://www.engineeringtoolbox.com/boiling-point-water-d_926.html, accessed 13 March 2014.
[24] Resistance thermometer, Wikipedia, 8 March 2014, http://en.wikipedia.org/wiki/Resistance_thermometer, accessed 13 March 2014.
[25] PT100 Platinum resistance thermometers, http://www.picotech.com/applications/pt100.html, accessed 13 March 2014.
[26] infra-red thermometer, Wikipedia, 7 January 2014, http://en.wikipedia.org/wiki/infra-red_thermometer, accessed 13 March 2014.
[27] Pyrometer, Wikipedia, 23 September 2013, http://en.wikipedia.org/wiki/Pyrometer, 13 March 2014.
[28] Black body radiation, Wikipedia, 20 February 2014, http://en.wikipedia.org/wiki/Black_body_radiation, accessed 13 March 2014.
[29] Bimetallic strip, Wikipedia, 5 March 2014, http://en.wikipedia.org/wiki/Bimetallic_strip, accessed 13 March 2014.
[30] Thermocouple, 12 March 2014, Wikipedia, http://en.wikipedia.org/wiki/Thermocouple, accessed 13 March.
[31] Liquid crystal thermometer, Wikipedia, 19 February 2014, http://en.wikipedia.org/wiki/Liquid_crystal_thermometer, accessed 13March 2014.
[32] Understanding weather, Met Office Education, http://www.metoffice.gov.uk/education/teachers/in-depth/understanding, accessed 13 March 2014.
[33] Layers of the atmosphere, US National weather Service, 22 July 2013, http://www.srh.noaa.gov/jetstream/atmos/layers.htm#ion, accessed 13 March 2014.
[34] Weather station, Wikipedia, 18 February 2014, http://en.wikipedia.org/wiki/Weather_station, accessed 13 March 2014.
[35] Min max thermometers, Weather station products, http://www.weather-station-products.co.uk/weather-featured-collections/min-max-temperature-thermometers?

page=1&results_per_page=100, accessed 13 March 2014.
[36] Humidity, Wikipedia, 10 March 2014,
http://en.wikipedia.org/wiki/Humidity, accessed 13 March 2014.
[37] Meehl G. A., et al. 2005, Sciencexpress, 10.1126/science.1106663,
http://www.nature.com/news/2005/050314/full/news050314-13.html, accessed 13 March 2014.
[38] Kiehl, J. T. & Kevin E. Trenberth, 1997: Earth's Annual Global Mean Energy Budget, Bulletin of the American Meteorological Society, 78 (2), 197–208.
[39] Meinshausen, M. et al., 2009: Greenhouse-gas emission targets for limiting global warming to 2°C, Nature, 458, 1158-1162.
[40] Keeling curve, Wikipedia, 10 March 2014,
http://en.wikipedia.org/wiki/Keeling_Curve, accessed 13 March 2014.
[41] Fifth assessment report, Intergovernmental Panel on Climate Change, http://www.ipcc.ch/index.htm#.UyG9mpWPPIU, accessed 13 March 2014.
[42] Stern Review, Wikipedia, 15 February 2014,
http://en.wikipedia.org/wiki/Stern_Review, accessed 13 March 2014.
[43] Economic impacts of climate change, Wikipedia, 31 December 2013, http://en.wikipedia.org/wiki/Economic_impacts_of_climate_change, accessed 13 March 2014.

Measuring the World, by John Austin

Measuring the World, by John Austin

9. Force

Force = that which causes an object or acts to cause an object to accelerate or decelerate

9.1 Introduction

In the *Philosophiæ Naturalis Principia Mathematica* (usually referred to as the *Principia*), published in 1687, Newton established that force is a fundamental action on an object, the tendency of which is to produce an acceleration according to the equation $F = ma$, where F is the force, m is the mass, and a is its acceleration[1]. An example is gravity. The apocryphal apple that fell from the tree observed by Newton[2], fell due to the force of gravity at an acceleration now measured (on average) as 9.80665 m s^{-2} or 32.174 ft s^{-2}. Prior to the work of Newton, it was considered that a force was necessary to keep an object moving even at constant velocity. Often, though, there is a hidden force (such as air resistance) which acts against any applied force. It took the perception of Newton to be the first to understand the true physics of forces. Strictly, the force is the rate of change of *momentum*, the product of mass and velocity (units of kg m s^{-1} or lb ft s^{-1}). For example, in the case of a rocket or comet, the mass decreases due to backwards expulsion of material, and this causes an acceleration.

Force is important in our daily lives as it determines for example the performance of our car engines needed for transport. This is often referred to as engine thrust, especially for aircraft engines. Of course we tend not to think about it, but when we 'weigh' ourselves, we are typically measuring the force due to gravity of our body mass. It's our mass that we normally intend to measure (Chapter 7), but weight is close enough as gravity doesn't vary appreciably. Forces are often spread over an area, and then the force per unit area is often significant. The force per unit area is the pressure, discussed in Chapter 10. Forces in every day life typically arise either from gravitation or the conservation of momentum. For example, for the rocket quoted above, if there are no gravitational forces acting, the rate of change of the momentum of the forward travelling section is equal and opposite to the rate of change of momentum of the backward travelling section.

Physics recognises the existence of only four forces: gravity, electromagnetism, the weak nuclear, and the strong nuclear[3]. It is thought that these four forces were initially unified at the time of the big bang, but as the universe cooled, so the forces became distinct. For example, gravity is thought to have separated from the other forces on the Planck time scale (Chapter 2) just 10^{-43} s after the big bang, the strong nuclear force separated after 10^{-35} s and the weak nuclear separated after 10^{-11} s. At this time the temperature would still have been some 10^{14} K[3]. Electromagnetism is itself a combination of electrical and magnetic forces, and is discussed in Chapter 12. The weak nuclear force is rarely encountered in normal life, and plays a role in radioactivity. The strong nuclear force is the force exerted between nucleons (neutrons and protons) in the atomic nucleus. One of the goals of physics is to unify the forces, by which is meant be able to describe all the forces in terms of a single equation. Unification of the forces then provides a means of understanding the physical processes which formed the universe. Work has already shown that electromagnetism and the weak nuclear can be treated as different forms of the same force. In the lower energy states present in the current universe, electromagnetism and the weak

Measuring the World, by John Austin

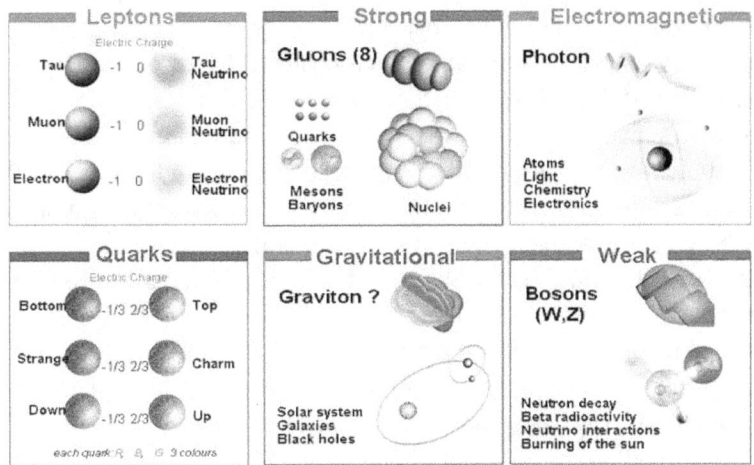

The four fundamental forces together with the exchange particles which realise those forces. Figure taken from [5]. The leptons and quarks are the building blocks of matter[6].

nuclear have become separate forces. Electromagnetism includes the forces between electrical charges and the forces on magnetic materials in a magnetic field. For these two forces, the photon, the particle equivalent of light (also known as electromagnetic) waves, is the exchange particle that provide the force. In other words, just like a rocket achieves an acceleration by expelling material, on an atomic scale, forces exist by the transfer of particles. In the case of electromagnetism, the particle is a photon. For the weak nuclear force, the exchange particles are the W and Z bosons[4].

While gravity seems like other forces at first analysis, it does have some unusual properties on a fundamental level. In particular, in most cases gravity is assumed to propagate as a force instantly. This would contradict other concepts, such as the existence of a finite speed limit for propagating signals. Also 'action at a distance' is a disturbing idea. Indeed it disturbed Newton. In other words, how can two objects that are physically separate affect each other? Our most up to date theory of gravity is Einstein's theory of General Relativity (GR) which treats gravity as a curvature of space[7]. So if there are no forces acting, an object would travel in a constant velocity in a straight line in curved space which would be perceived as an acceleration under gravity. GR resolves the issue of the finite velocity of propagation of a signal by limiting the signal speed to c, the velocity of light in a vacuum. However, the most comprehensive theories of gravitation seek the exchange particles, called gravitons[8]. Because gravity has an infinite range (like electromagnetism), gravitons are expected to be massless, like the photon. Experiments have been set up to find these elusive particles, but to date (October 2014), the graviton or gravitational waves have not been found.

In this Chapter, we discuss gravity and other forces, as traditionally recognised, and delay the discussion of electromagnetic forces to chapter 12. The strong and weak nuclear forces are complex and do not impinge directly on our daily experiences, and are therefore not considered further. In § 9.2 and 9.3 the units of force and the conversion of units are given, while in § 9.4 some rough magnitudes of everyday forces are given. In § 9.5 - 9.8 gravity is discussed, focussing on the Earth in § 9.6 – 9.7. Tides arise from gravitational forces and are discussed in § 9.9. The remainder of the Chapter focusses on other forces – friction, rotational forces and finally hardness and surface tension.

9.2 Units of Force

In the fps system, forces are measured in poundals (abbreviation pdl)[9]. With masses in pounds, $F = ma$ will yield accelerations in ft s^{-2} when F is in poundals. The poundal is often confined to engineering applications, although why they haven't adopted the metric system is a complete mystery to me. Often engineers measure forces in pounds force with the implicit understanding that to get to poundals, you need to multiply by the acceleration due to gravity (32.2 ft s^{-2}). Or, put another way, you can work with the forces in pounds, but need to multiply by g at the end of the calculation to recover the correct acceleration.

In the cgs system, the unit of force is the dyne (dyn), which is rather small for most practical purposes as it produces an acceleration of only 1 cm^{-2} in a small weight of 1 g. The SI system is more practical because of its much larger units. The unit of force is the newton (N) which causes an acceleration of 1 m s^{-2} in a mass of 1 kg. Although still slightly small for many purposes it can of course be used with metric pre-multipliers.

9.3 Conversion of Units

Because of the way the units are derived in the metric system, 1 dyn is a small force: 1 N = 1000 g x 100 cm s^{-2} = 10^5 g cm s^{-2} = 10^5 dyn. 1 pdl = 1 lb x 1 ft s^{-2} = 0.45436 kg x 0.3048 m s^{-2} = 0.136 N. So, to convert from newtons to poundals multiply by 7 (4% error) and an approximate conversion from poundals to newtons is to divide by 7. Another conversion, e.g. Used for thrust is 1 pound force = 0.45436 kg x 9.8 m s^{-2} = 4.36 N. So, an approximate conversion is to divide the thrust in newtons by 4 and subtract 10% (error 1%). To convert pounds to newtons, multiply by 4 and add 10% (error 1%).

9.4 Approximate Forces

Body weight is of order 700 N for the average adult. So a very strong person (e.g. An international weightlifter) is capable of exerting a force of at least 1000 N. Vehicle thrusts may exceed 200 kN for a jet engine and up to 8,000 kN for a rocket engine[10].

I would guess that a small car would be several kN, but I have not succeeded in finding precise information directly. Engineers are more interested in telling us the brake horse power (sic) or the amount of CO_2 generated per kilogramme of fuel. Nonetheless, with a bit of detective work, we can calculate this.

It turns out that the venerable Reliant Robin, the cult non-car car, does 0-100 km h^{-1} in about 16 s[11], an 'acceleration' of 100,000/(3600x16) = 1.74 m s^{-2}. Its mass is 436 kg[12], but you have to include say a 64 kg person to bring the total mass to a round 500 kg. *F = ma* gives *F = 500x1.74*, or barely 1 kN (!), assuming that rolling friction is negligible. The force is roughly that exerted by a reasonable weightlifter.

9.5 Understanding Gravity

As long as as 300 BC Aristotle is recorded as believing that heavy objects fell faster than light objects[13]. Eventually by the early 1600s, Galileo is thought to have demonstrated that all objects fell at the same rate, but the story that he experimented with objects dropped from the leaning tower of Pisa appears like many historical accounts of science of that era to be apocryphal[14]. A modern version to show that all objects fall at the same speed is to place a feather and a heavy object together in an evacuated tube and turn it around. However, if the air is pumped back in, the heavy object will indeed fall faster due to lower air resistance[15].

During the 17th Century, Kepler also put forward a set of laws which determined the motion of the planets around the sun. These laws pointed out that the planets orbited the sun in elliptical orbits, with the sun at one focus of the ellipse. Following a thorough mathematical analysis of these laws, Newton was able to provide an explanation for the gravity determining the motion of the planets around the sun[16]. Moreover, he reasoned that the

Measuring the World, by John Austin

basic law of gravity also determined the force of gravity on Earth, finally providing an explanation for why objects, such as the apocryphal apple fell from the tree, towards Earth. Newton's law of gravitation has been found to be accurate to high precision. It was not until 1915 that Newton's law was finally superseded by the theory of General Relativity[7], which was needed to explain some of the observed discrepancies in Newtonian theory. Because these discrepancies are unimportant in the vast majority of situations encountered by people, Newton's law of gravity is still used almost exclusively. In the theory, two masses m_1 and m_2 attract each other with a force $F = Gm_1m_2/r^2$, where G is the Universal Gravitational constant (Chapter 2) and r is the distance between the centres of mass of the two bodies. Note that the law is universal: it works anywhere on Earth, it determines the planetary orbits, and even the forces between galaxies. The only exceptions are adjustments that are required by the more accurate GR theory. As noted in § 9.1, GR theory treats gravity as a curvature of space itself. GR explains the discrepancies of Newtonian gravitation, including the observed bending of light rays by the sun and anomalies in the orbit of the planet Mercury.

    One of the conclusions of GR is that acceleration and gravity are equivalent forces as far as an observed is concerned. Suppose an observer is in a rocket ship which is accelerating forwards. If the observer is standing on the base, he would feel a downward force as if he were standing still on a planet surface. A space ship that is moving without acceleration, but rotating rapidly would also cause a force (the centripetal force) on the inner rim. If the radius of the craft is r, the centripetal force is mv^2/r, which equals the equivalent weight mg. So, $rg = v^2$. If $g = 10$ m s^{-2} (i.e. near enough the same as Earth's gravity) and $r = 100$ m, then $v = 30$ m s^{-1} approximately, giving one rotation in 20 s. In other words, if we had the technology to go on long space journeys, we could simulate Earth gravity by rotating the spacecraft.

Measuring the World, by John Austin

9.6 Measuring Earth's Gravity and Earth's Rotation

A simple way to measure the acceleration due to gravity, g, is to use a pendulum. Galileo apparently started thinking about this problem during an especially uninspiring sermon in about 1582[17]. However, Galileo seems to attract a number of apocryphal stories, so don't be surprised if this one doesn't turn out to be true either! A pendulum consists of a weight which is allowed to oscillate about a fixed support. We consider the case in which the weight is attached to a string much longer than the size of the weight. For small oscillations, less than about 5° on either side of the vertical, the time for a complete oscillation is $2\pi(l/g)^{½}$. The length of the pendulum is l, and g is the acceleration due to gravity. In other words, the time is independent of the weight, and is independent of the size of the oscillation (for small oscillations). This was Galileo's observation: the period remained the same for a swinging *objet d'art* above the pulpit, even as the amplitudes became very small. Under controlled conditions, the length of the pendulum can be determined to a precision of 0.1% with no special equipment, and the time for 1000, say, for l = 1 m approximately, can give a good precision measurement of g (error 0.1%) with hardly any effort. Even a pendulum can show the variation of surface values of g, from about 9.76 m s^{-2} at the equator to 9.83 m s^{-2} at the poles[18]. Also, g decreases with height and reduces by about 10% by 400 km altitude, the height near which most satellites operate.

Specialised equipment can make very high precision measurements, either using the change in length of a spring due to a test mass, or a direct timing of the free fall of an object in controlled conditions[19]. Quoted uncertainties of these measurements are less than 2x10^{-8} m s^{-2}. One of the advantages of knowing g to high precision, is that rocks of low gravity can be detected. These are likely to be porous, and are a potential source of oil[20].

Foucault pendulum at the Museu de les Ciencias Principe Felipe (science museum) in Valencia Spain. Author: Manuel M. Vicente (Spain) 17 September 2006[22]. Science and art is combined in a beautiful picture! The little balls around the rim of the pendulum base are successively knocked down by the pendulum and are used to show its progress. Just think: some poor soul has to mount these all again properly at the end of the pendulum day, possibly running the gauntlet of trying to avoid being hit by the pendulum!

Another important use of the pendulum was established by Foucault in 1851[21], and which now bears his name. He attached a 67 m long steel wire to a ceiling support that was free to rotate. To the other he attached a 28 kg bob. An experiment was carried out in the Panthéon in Paris in which the pendulum was set swinging. After several hours, the pendulum appeared to have changed direction relative to the ground. Rather, with a low friction support, the pendulum continued to oscillate in the same plane (by Newtonian dynamics) but the Earth rotated on its axis. In fact the pendulum covers a complete circle in 24 hours/$\sin \vartheta$, where ϑ is the latitude. In principle this could be used to determine latitude, if using the sun and stars was not your thing. Of course at the equator, $\sin \vartheta = 0$, and the pendulum doesn't change its plane of rotation relative to the Earth, but at the poles where $\sin \vartheta = 1.0$, the pendulum completes a circle exactly once in 24 hours. When I was in my early years at high school, the 6th form boys had set up a Foucault pendulum in one of the high school stairwells. I was most fascinated and it was perhaps one of the factors which inspired me to do research in atmospheric sciences. Meteorological processes of course depend critically on the rotation of the Earth. Unfortunately, by the time I was in the 6th form, our teachers didn't let me set up the experiment, saying that it was too difficult!

9.7 The mass of the Earth

It may seem remarkable at first that we can get a surprisingly high precision measurement of the mass of the Earth, but the Newtonian law of gravitation enables us to do so. The force between a mass m on the Earth's surface and the Earth itself with mass M is GmM/R^2 by Newton's law, where R is the radius of the Earth. But, this is just the gravitational force on the object, mg, so: $GmM/R^2 = mg$

$\rightarrow M = gR^2/G = 9.80665 \times (6.371 \times 10^6)^2/(6.6738 \times 10^{-11})$

$\rightarrow M = 5.964 \times 10^{24}$ kg

The above calculation depends in particular on an accurate measurement of G. The precision of the Earth mass measurement is almost the same as that of G itself, 1 part in 10^4. In fact the above simple calculation is not quite accurate, as you need to take account of the centripetal force on the rotating Earth, which lowers the observed value of g. The accepted mass of the Earth, then, is slightly higher (0.14%) at 5.97219×10^{24} kg[23].

There have been some truly exquisite experiments which have succeeded in measuring the tiny displacements of μg-sized masses[24]. Experiments have revealed the reliability of the law of gravitation over distances from 10 μm to astronomical sizes[25]. Nonetheless, G is not known to high precision compared with other fundamental constants (Chapter 2), and this relatively poor precision may be masking any inaccuracy in the law of gravitation, which may not be an inverse square law with distance throughout the observed range. Therefore, physicists continually probe the law of gravitation to see whether an alternative functional form might better represent the observed forces between masses. To date, no revisions of the inverse square have been deemed necessary.

9.8 Gravitational forces on satellites and spacecraft

To place a satellite in orbit implies the need to impart significant energies to the payload using powerful rockets. This is considered routine by many countries, and as of January 2013 there were about 3600 operating satellites in orbit around the Earth[26], out of a total of 6600 launched so far. Orbits are classified as geosynchronous or polar. The former has a rotational period of 24 hours, to match that of the Earth. They are used for continuous observation of the surface such as in military spying operations, or for weather observation. For the orbit to be stable it needs to be high altitude (35,700 km) and the satellite needs to be placed over the Equator. Such satellites give a relatively poor view of high latitude regions, so for improved scanning, polar orbiters are used. The orbit can have any angle desired with respect to the Equator, but the nearer the angle is to 90 degrees, the nearer the overpass of the satellite to the poles. Often, for meteorological purposes, the orbit is sun-synchronous. That is, the sub-satellite point crosses a given latitude at the same local time each day. Polar orbiters are usually much lower in altitude than geosynchronous satellites, only a few hundred kilometres, giving about 14 orbits per day. They can be used for observations of the whole Earth over a 24 hour period, and with two passes per day over or close to a given location. These satellites are used for research, navigation (GPS) and a wide range of other purposes. To keep a satellite in orbit, its speed along the orbit is sufficient for the centripetal and gravitational forces to balance:

$$mv^2/(R + h) = GmM/(R + h)^2 \rightarrow v^2 = GM/(R+h)$$

R is Earth's radius, h is the height and m is the mass of the satellite, G is the gravitational constant and M is Earth's mass. But, from § 9.7, $GM = R^2g \rightarrow v^2 = R^2g/(R+h)$. For the polar orbiter, $h = 300 \times 10^3$ m $\rightarrow v^2 = (6.37 \times 10^6)^2 \times 9.81/6.67 \times 10^6 = 6 \times 10^7 \rightarrow v = 7725$ m s^{-1} or about *27,800* km h^{-1} *(17,300 mph).*

Getting into orbit is the first step towards escaping Earth's gravity entirely. From the satellite altitude of 300 km, to escape from the Earth it needs to have kinetic energy which exceeds the gravitational potential. Hence, ½mv^2 > $GmM/(R+h)$.

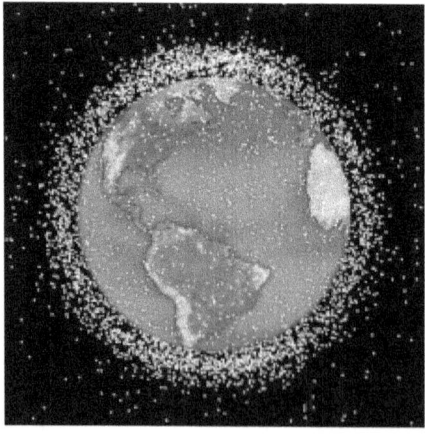

More than 500,000 pieces of "space junk" are tracked by NASA as they have the potential to cause serious damage. Image by NASA[27].

Using $GM = R^2g$ again, $v^2 > 2R^2g/(R+h)$. i.e., the velocity has to be a factor √2 greater than before, or 39,300 km h^{-1} (*24,400 mph*).

The launching of satellites has perhaps been too successful in some ways. There are now over 500,000 pieces of debris from the breakup of satellites and other causes[27] which are currently orbiting the Earth. Such pieces, some of which are only a few cm across, have the damage to case severe damage as they are travelling at such high speeds, as much as 27,000 km h^{-1}. China was accordingly criticised in 2010[28] for using a missile to destroy one of its defunct satellites. At least one satellite may already have been damaged by the debris from the break up[29]. The only way that the resulting thousands of particles could become safe is by eventual burn up in the atmosphere as their orbits decay.

It is possible to turn gravitational forces to our advantage using the 'gravity-assist' method[30, 31], which works on the principle of accelerating a probe as it approaches a large object such as a planet. For this to work, the probe needs to approach the planet in the same direction as its orbital motion. The probe acquires a significant fraction of the velocity of the planet to add to its approach velocity. So, although the probe slows down after approach, there is a net gain in velocity by the probe, and in fact a minuscule slowing down of the planet! The method has proved to be an invaluable way of exploring the solar system and the Voyager programme has been a spectacular success[32,33]. The voyager spacecrafts were launched in 1977 to take advantage of a once in 176 year alignment of the outer planets. The gravity assist method was essential to raise the velocity of the spacecraft to allow it to visit the planets Jupiter, Saturn and Uranus, while Voyager 2 also visited Neptune. The main impetus was supplied by Jupiter, which raised the velocity of Voyager 1 from 49,000 km h^{-1} to 86,000 km h^{-1}, (relative to the sun). Voyager 2's velocity roughly doubled, from 38,000 to 75,600.

Measuring the World, by John Austin

As of March 2014 Voyager 1 and 2 had velocities 17.0 and 15.4 km s^{-1} relative to the sun, and were at distances of 19.0 and 15.7 billion km respectively[33,34]. [34 is quite confusing, though, as the units are not written with the proper abbreviations.] To put the speeds into perspective, the Voyager 1 speed is about 60 millionths of the speed of light, so it will take thousands of years to reach the next star system.

9.9 Tidal Forces

Gravitational forces are also responsible for ocean tides on earth. The net tide arises from the combined gravitational pull of the sun and moon and the rotation of the Earth[35]. The time between peak tides is 12.4206 h so there are two tides per day, and the times gradually shift later in the day. This period is almost 12 hours, but is slightly more (0.4206 h or 25 min 14 s) due to the time taken for the moon to rotate in its orbit sufficiently to be over the same point on earth. The tide is described as a 'neap tide' when the sun and moon are pulling at right angles to one another. A 'spring tide' occurs when the sun and moon are pulling along the same line of action.

The tidal force is the difference in the force of gravity across the diameter of the Earth. The force of gravity is $F = Gm_1m_2/r^2$ where m_1 and m_2 are the masses, and r is the orbital radius. So the difference is the magnitude of $dF/dr = 2Gm_1m_2/r^3$ (simple differentiation) multiplied by the Earth's diameter. Since we are only interested in ratios, we need to compare $F' = m_1/r^3$ where m_1 is the mass of the sun or moon respectively. For the sun, $m_1 = 1.9891 \times 10^{30}$ kg, $r = 149.6$ million km $= 1.496 \times 10^{11}$ m[36] and $F' = 1.9891 \times 10^{30}/(1.496 \times 10^{11})^3) = 0.5941 \times 10^{-3}$. For the moon $m_1 = 7.3477 \times 10^{22}$ kg, $r = 384400$ km $= 3.844 \times 10^8$ m[37], and $F' = 7.3477 \times 10^{22}/(3.844 \times 10^8)^3 = 1.293 \times 10^{-3}$. Therefore, although the sun is much more massive than the moon, because of its distance, the solar tide is only 0.5941/1.293 ≈ 46% of that of the moon (see also [38], which gives 44% for the ratio). This fraction varies according to the distance from the sun and variations in the moon's orbit, which together will give rise to different height tides.

9.10 Frictional Forces

While it is customary to think of friction as a force to be reduced as much as possible, its presence is not always a disadvantage (e.g. In braking a fast-moving vehicle). Frictional forces occur when two surfaces slide over each other and there is a load on the surfaces. Static friction occurs before objects actually start to slide, and sliding friction, which is smaller in magnitude, occurs when there is motion between the surfaces. The frictional force divided by the load on the surface is known as the coefficient of friction (C_f) and is independent of the load to a good approximation. The C_f value for steel on steel can be as high as 0.8, but for lubricated steel such as a train on its tracks the value is nearer 0.16[39]. Rolling friction is about 1% of sliding friction. The C_f value for wood on dry metal is 0.2 – 0.6[40]. Consider a train moving along tracks. The mass of the train we take as 200 t, C_f = 0.15 for sliding friction and 0.02 for rolling friction. The forces are weight (load) 200x9.8 ≈ 2 MN, sliding friction 0.3 MN and rolling friction 4 kN. A train, then has a small resistance to motion which is energetically efficient but it is also a problem, as it cannot stop quickly. For the above example, if the brakes are engaged, the deceleration is the frictional force divided by the mass = $0.3 \times 10^6 / 2 \times 10^5$ = 1.5 ms^{-2}. This is consistent with the estimate for the deceleration given by [41]. A train running at 50 m s^{-1} (180 km h^{-1}) will come to a halt in 50/1.5 = 33.3 s. The distance travelled, x, is given by the equation $v^2 = 2ax$, where v is the initial speed and a is the deceleration. This gives x = 2500/3 = 830 m approximately. This is a long way (over half a mile in fps) and illustrates why stalling your car on a level crossing is not a clever idea! The energetic efficiency of a fast moving train is to be balanced against the difficulty of accelerating and decelerating.

The transport characteristics of a sports car are the opposite of that of a train. Sports cars are provided with wide tyres which increase traction (and friction) for acceleration and deceleration. Braking distances are important of course for ordinary cars as well. When I was training for my driving examination in about 1980, my instructor gave me the formula (fps!) for the minimum stopping distance. We were supposed to memorise the distance at specific speeds. The relationship was $s = v + v^2/20$. The minimum stopping distance is s (in feet) for the speed, v, in mph. So, for 30 mph, $s = 75$ ft and for $v = 70$ mph, $s = 315$ ft. In the expression, there is a term equal to v, the distance travelled during the time taken (estimated to be about ¾ s) to get your brain to engage your right foot. $v^2/20$ is the deceleration term which shows that taking kinetic energy from the car is proportional to distance, as confirmed by the earlier simple dynamics of a constant decelerating force for the train. If the car were moving at the same speed as the train, $v = 110$ mph, the minimum stopping distance would be 12100/20 = 605 ft excluding 'thinking time' which has also been excluded in the case of the train. The distance is about 180 m, a factor of 4 less than the distance for the train. So, rubber tyres on asphalt makes a much more effective brake than steel-on-steel. $C_f \approx 0.7$ in fact[40]. Hence the improved braking is in the ratio of the respective values of C_f. Incidentally, I never remembered the numbers at different speeds, I merely did the calculation ($v + v^2/20$) mentally when asked by the examiner. This formula, is now decades old and hasn't been updated[42]. It may seem strange that 50 years of improvements in brakes and tyres haven't reduced the 'minimum' stopping distance. However, this reflects the fundamental aspect of braking: the maximum braking force is the road friction and there has been a tendency towards reducing friction rather than increasing it, in the process of making cars more energetically efficient. Interestingly, in the driving instructions for some US states, you are supposed to allow a 2 s gap from the car in front, but this is clearly inconsistent with the quadratic form of the correct physics.

Observations indicate that if the density of traffic exceeds about 40-50 vehicles per mile (so cars would be about 1 s apart), 'flow breakdown' occurs[43]. In this case, a wave occurs long the line of traffic as drivers brake to avoid collisions. The mean speed of the traffic then drops precipitously. The flow can only be resumed slowly from the front of the line once large gaps are re-established. The theory of traffic flow is a fascinating practical problem which has attracted considerable mathematical analysis[43].

Ice skating is another low friction environment, and skates are designed for their particular task[44]. For example, in speed skating, the skate blades are straight, flat and thin, minimising energy use. Momentum is gained with quick running-like strides and maintained with long strokes of the legs and with a wide swing of one or both arms. Changing direction or speed is slow. Hockey skates have blades that are curved at each end, enabling sharp turns to be made while chasing after the puck. Figure skates have blades with several teeth at the front for gripping the ice to aid the completion of special gymnastic manoeuvres.

9.11 Rotational forces

As noted in § 9.1, linear momentum (i.e. momentum in a fixed direction) is conserved in the absence of forces. The angular equivalent is the conservation of angular momentum, which is the product of 'moment of inertia', I[45] and the angular velocity, ω. The angular velocity is measured in radians per second ($rad\ s^{-1}$) and an angle of 2π radians is the same as one rotation or $360°$ of angle, where π has the usual value 3.14159.... So, 1 revolution per second is the same as an angular velocity 2π rad s^{-1}. The moment of inertia is difficult to calculate for many objects, and is always specific to a given axis of rotation. To do the calculation requires a sum (integral) over all the particles in the object of the product of their mass and the distance squared from the axis. The calculation usually needs the integral calculus, but is easy for some simple objects[46]. Consider for example a ring of radius r and mass m. Take the axis passing through the centre of the ring at right angles to its plane. Then every particle is the same distance from the axis, so the moment of inertia is $I = mr^2$. The equivalent of the rate of change of linear momentum is C = *the rate of change of* $I\omega$, where C is the couple acting on the axis, i.e., the force times the distance from the axis.

There are several practical or well-known examples aspects of these principles. Ice dancers frequently perform a spin on the ice. He or she can start with arms extended, hence maximising I about the vertical axis. Then, if a couple is exerted on the ice then the dancer will start spinning. The pulling in of the arms reduces I and hence the rotation rate ω needs to increase to conserve $I\omega$. In other words, the conservation of angular momentum ensures that the spin rate increases. Over its history, figure skating and ice dancing have made extensive use of the laws of dynamics, such as the conservation of angular momentum in a relatively friction-free environment and these advances have spawned a large number of different terms that the layman can scarcely comprehend e.g. [45]. In addition, other moves that one might expect to see, do not occur in competition. For example, I wondered why I never saw forward or backward somersaults which I thought might be possible by a gifted performer. It turns out that these have been banned from competition[47], presumably on the

Measuring the World, by John Austin

grounds of safety.

An entertaining demonstration of the physics of rotation can be obtained with an empty shoe box. Like all 3-D objects, a shoe box has three possible axes of rotation. If you spin the box around an axis through its length, the box spins smoothly. If you spin the box about an axis perpendicular to the top of the box, the box also spins smoothly. However, if you try spinning the box around the third and final axis, it twists out of the spinning plane[48]. In the jargon of physics, 3-D objects are unstable to rotation about the axis of the middle moment of inertia, a point that can be demonstrated using advanced mathematics [49, although this is a bit messy: a more elegant way is using Generalised Hamiltonian Dynamics, if I recall correctly, but I don't have a reference any more and my old lecture notes are hidden in the loft!]. Lest you think that this is all very entertaining but not relevant to the real world, consider how you have to stabilise a helicopter to avoid it smashing itself to pieces.

Conservation of angular momentum is critical for the stability of the helicopter[50]. The rotation of the main rotor blades provides lift according to Bernoulli's theorem in a fluid[51], in this case air. The angle of the rotors to the horizontal and their speed of rotation determine the lift and speed of the helicopter. In addition, the rotors exert an undesirable horizontal couple which spins the machine in the opposite direction as the rotors. In the absence of any counteracting couple, the aircraft would spin out of control and crash. Stability is ensured by the tail rotor, which like any aircraft provides a force perpendicular to the tail rotor blades. Because the tail rotor is a long distance from the centre of gravity it provides a relatively large couple (force x distance from the centre of gravity) which balances the couple (a larger force x a small distance from the c.g.) induced by the main rotor. Fun though they are to be in, it seems to me that there are so many things that can go wrong with a helicopter[50] that it's a wonder that we put so much faith in them.

Measuring the World, by John Austin

Gyroscopes[52] take advantage of the conservation of angular momentum to detect a change in direction or acceleration of objects, especially aircraft. They have high spin rates using a device for each of the three spatial dimensions, and are used for inertial navigation by detecting the tilt of their mounting. Gyroscopes need a high moment of inertia for extra sensitivity, and this can be achieved using tungsten, which is one of the densest materials available at 19,300 kg m^{-3} at 0 °C.

On a somewhat larger scale, the motion of the Earth is influenced by coupling forces from the Sun and Moon. Applied to the axis of rotation, these cause the Earth to precess slightly, similar to a spinning top. In other words, the axis of rotation doesn't point to a fixed point in space, but gradually traces a small circle over a period of 26,000 years[53]. There is an additional precession of the Earth's orbit around the sun, which contributes to changes in the Earth's climate, the so-called Milankovitch cycles[54]. Over a period of 19,000 -- 23,000 years, the tilt of the Earth's axis varies between 22.1° and 24.5° and is currently about 23.4° [55]. The larger the tilt, the larger the differences between the seasons, so at the upper end of the range, summers would be hotter and winters colder than when the Earth's tilt is nearer the lower end of its range.

On a trivial level, understanding of rotational forces can be used to distinguish between a hard boiled and a fresh egg. Place the egg horizontally on a table and manually spin it as fast as possible. A hard boiled egg will spin faster and more steadily. This is because the moment of inertia is lower and constant, so for a given angular momentum the rotation rate will be higher. For a fresh egg, the yolk is displaced from the centre during spinning, and it is more dense than the egg white. Consequently, the moment of inertia is higher, and varies as the yolk moves around inside the spinning egg, so the spinning occurs at a slower rate and unevenly. Of course the way most people determine whether an egg is fresh or not is to crack it open, but once in your life you may need to do this experiment!

9.12 Measuring Spin Rates

Spin rates can be measured with a tachometer[56] and the units in everyday use are revolutions per minute (*RPM*). A magnet attached to a spinning object induces an electric field (Chapter 12). The voltage is proportional to the speed. This device is used in vehicles to determine when to change gear automatically, or simply to diagnose engine revolutions. In CD/DVD readers and writers, tachometers are used to control the rotation rate of the disc. Tachometers are also used in computers to sample data hard drives at set speeds.

Exotic physical properties can be exploited to determine rotation rate. For example, Helium (He-4) becomes a superfluid at temperatures lower than 2 K (-271 ºC). A superfluid flows without viscosity, the fluid version of friction (§ 9.13). A pressure difference between two tubes connected to He-4 causes oscillations which occur as sound waves, and can be detected with headphones in contact with the liquid. The method can be used to detect changes in the Earth's rotation rate[57].

9.13 Other Forces

Other forces are important in our daily lives and we scarcely give them a second thought. One of these, hardness, does not usually have any units but is measured with Mohs scale of mineral hardness[58]. It is used to determine the impact of one material on another, e.g. Diamond on glass. Hardness has mostly industrial applications, indicating whether one material can scratch or cut another. Approximate values for minerals are: Talcum 1; Gypsum 2; Calcite 3; Fluorite 4; Apatite 5; Feldspar 6; Quartz 7;

Topaz 8; Corundum 9; Diamond 10. The higher the rating (hardness), the more the mineral will scratch a material of lower hardness. Other ratings for common materials are: Graphite (pencil lead) 1.5; Fingernail 2.2 – 2.5; Copper coin 3.2 – 3.5; Window glass 5.5; Knife blade 5.5; unglazed porcelain 7.0. Accurate testing of materials is carried out with a sclerometer[59] which measures the force needed to scratch the test material with a diamond or Borazon-tipped tool.

Viscosity is the fluid analogy of friction (§ 9.9). The force arises from the attraction of molecules or their collisions, and opposes the flow of the fluid. For example, viscosity needs to be considered in the flow of natural gas along a pipeline. Detailed mathematics[60] shows that the rate of flow of a fluid is inversely proportional to the viscosity of the fluid and strongly dependent on the radius of the pipe. This theory covers a very broad range of topics including lung alveoli, hypodermic syringe, flow in a water pipe as well as the gas flow mentioned. The flow rate increases as the fourth power of the pipe radius, so that a doubling in radius increases the flow rate by a factor of 16. This is because the viscosity dominates the fluid behaviour close to the vessel walls. In a liquid, viscosity arises from attractive forces between molecules due to their electronic charges. As the temperature increases, the molecules move further apart. The electrical attractions then become smaller, and hence the viscosity generally decreases with temperature. In a gas, the viscosity arises mostly from collisions between molecules. As the temperature increases, molecular velocities increase, so the rate of collisions, and the viscosity also increase[61].

Surface tension allows bubbles to form in liquids as well as other phenomena, including the support of objects denser than water[62]. For example, when a liquid begins to boil, due to heating from below, bubbles of vapour form spontaneously within the liquid, travel up the container, and dissipate as vapour. The surface tension is the force per unit length at right angles to a line

Measuring the World, by John Austin

drawn in the surface. A bubble requires excess pressure due to surface tension of $2\gamma/r$ where γ is surface tension and r is the bubble radius, but a soap bubble has an inner and outer surface, so requires twice the pressure difference. This implies that a very large force is needed to get the bubble started. In fact the function is infinite ("singular") for $r = 0$, and the pressure gets smaller as the bubble gets larger. In practice, bubbles form around liquid impurities which may have sizes of order a few μm or more. This sets a limit for the excess internal bubble pressure at about an atmosphere or less. However, for pure water with few impurities to start the process of bubble formation, the water becomes superheated and eventually boils explosively, as shown in this entertaining video[63].

Insects can float on water due to surface tension. A typical housefly of size 5-7 mm would be subject to a surface tension $2 \times 7.3 \times 10^{-2} \times 6 \times 10^{-3} \cos \alpha \approx 8.8 \times 10^{-4} \cos \alpha$ N[62], where α is the angle between the surface and the local vertical. Taking $\cos \alpha = 0.5$, surface tension is sufficient to support a mass of 40 mg, over three times the 12 mg mass of a typical housefly[64]. So, the fly can walk on water! For humans in water, surface tension is a negligible force, of order 0.1 N, but to an insect, water has a very different appearance, rather like treacle to us, at least I assume so: it's often very difficult getting basic information about common substances. The main point is that if we could find a liquid with surface tension 3.5×10^5 N m^{-1}, this would feel the same to us as water does to a housefly. This large difference in perception of liquid surface tension is primarily due to scaling: the ratio of weight to surface tension has the dimension of a length so small objects (insects) have a higher chance of being supported than large objects. Addition of soap reduces surface tension. So, if you have an annoying insect floating in your toilet bowl, add some liquid soap (away from the prying eyes of animal cruelty organisations!) and the animal will sink and drown.

Measuring the World, by John Austin

References

[1] Newton's laws of motion, Wikipedia, 14 March 2014, http://en.wikipedia.org/wiki/Newton's_laws_of_motion, accessed 14 March 2014.
[2] Isaac Newton, The apple incident, Wikipedia, 8 March 2014, http://en.wikipedia.org/wiki/Isaac_Newton#Apple_incident, accessed 14 March 2014.
[3] The fundamental forces of nature, Web Syllabus, Astronomy 162, Dept. Physics and Astronomy, University of Tennessee, http://csep10.phys.utk.edu/astr162/lect/cosmology/forces.html, accessed 14 March 2014.
[4] Weak interaction, Wikipedia, 23 February 2014, http://en.wikipedia.org/wiki/Weak_interaction, accessed 14 March 2014.
[5] Four fundamental forces, by Ananth, 19 August 2010, http://countinfinity.blogspot.co.uk/2010/08/four-fundamental-forces.html, accessed 14 March 2014.
[6] Lepton, Wikipedia, 10 February 2014, http://en.wikipedia.org/wiki/Lepton, accessed 14 March 2014.
[7] General relativity, Wikipedia, 12 March 2014, http://en.wikipedia.org/wiki/General_relativity, accessed 14 March 2014.
[8] Graviton, Wikipedia, 8 February 2014, http://en.wikipedia.org/wiki/Graviton, accessed 14 March 2014.
[9] Poundal, Wikipedia, 28 December 2013, http://en.wikipedia.org/wiki/Poundal, accessed 14 March 2014.
[10] Jet engine, Wikipedia, 22 December 2013, http://en.wikipedia.org/wiki/Jet_engine, accessed 14 March 2014.
[11] Robin, Reliant Owners Club, 2010, http://www.reliantownersclub.co.uk/robin.html, accessed 14 March 2014.
[12] Specifications: 1981 Reliant Robin 850, Unique cars and parts, http://www.uniquecarsandparts.com.au/specifications/reliant/1981_reliant_robin_850.htm, accessed 14 March 2014.
[13] Gravitation, Wikipedia, 2 March 2014, http://en.wikipedia.org/wiki/Gravitation, accessed 14 March 2014.
[14] Tall tales, P. Ball, Nature News, 17 June 2005, http://www.nature.com/news/2005/050613/full/news050613-10.html,

Measuring the World, by John Austin

accessed 14 March 2014.
[15] Feather and ball being dropped in vacuum, 14 June 2008, http://www.youtube.com/watch?v=_XJcZ-KoL9o, accessed 14 March 2014.
[16] Kepler's laws of planetary motion, Wikipedia, 13 March 2014, http://en.wikipedia.org/wiki/Kepler's_laws_of_planetary_motion, accessed 14 March 2014.
[17] Pendulum, Wikipedia, 10 March 2014, http://en.wikipedia.org/wiki/Pendulum, accessed 14 March 2014.
[18] Gravity of Earth, Wikipedia, 14 February 2014, http://en.wikipedia.org/wiki/Gravity_of_Earth, accessed 14 March 2014.
[19] Gravimetry, 18 April 2013, Wikipedia, http://en.wikipedia.org/wiki/Gravimetry, accessed 14 March 20014.
[20] Gravimetry, magnetics and electromagnetics, RWE Technology and Innovations, http://www.rwe.com/web/cms/en/1773366/rwe-dea/know-how/exploration/gravimetry-magnetics-electromagnetics/, accessed 14 March 2014.
[21] Foucault pendulum, Wikipedia, 13 March 2014, http://en.wikipedia.org/wiki/Foucault_pendulum, accessed 15 March 2014.
[22] Foucault pendulum at the C.A.C. In Valencia (Spain), Wikimedia, Manuel M. Vicente, 17 September 2006 http://commons.wikimedia.org/wiki/File:P%C3%A9ndulo_de_Foucault_(M._Ci%C3%A8ncies_Valencia)_01.jpg, accessed 15 March 2014.
[23] Solar system exploration, Earth facts and figures, NASA, 25 October 2013, http://solarsystem.nasa.gov/planets/profile.cfm?Object=Earth&Display=Facts, accessed 15 March 2014.
[24] Gravity test constrains new forces, Physics World.com, News, 5 March 2008, http://physicsworld.com/cws/article/news/2008/mar/05/gravity-test-constrains-new-forces, accessed 15 March 2014.
[25] Testing the law of universal gravitation, Learner Teacher resources, Physics for the 21st century, Annenberg Foundation 2014, http://www.learner.org/courses/physics/unit/text.html?unit=3&secNum=5, accessed 15 March 2014.
[26] Satellite, Wikipedia, 5 March 2014,

http://en.wikipedia.org/wiki/Satellite, accessed 15 March 2014.
[27] Space debris and human spacecraft, NASA, 27 September 2013, http://www.nasa.gov/mission_pages/station/news/orbital_debris.html, accessed 15 March 2014.
[28] US report claims china shoots down its own satellite, China Daily, 19 July 2010, http://www.chinadaily.com.cn/world/2010-07/19/content_10121179.htm, accessed 15 March 2014.
[29] Russian spacecraft hit by 'space junk' from destroyed Chinese spacecraft, RT News, 9 March 2013, http://rt.com/news/russian-satellite-collide-chinese-044/, accessed 15 March 2014.
[30] A gravity assist primer, NASA JPL, California Institute of Technology, http://www2.jpl.nasa.gov/basics/grav/primer.php, accessed 15 March 2014.
[31] Gravity assist, Wikipedia, 14 March 2014, http://en.wikipedia.org/wiki/Gravity_assist, accessed 15 March 2015.
[32] Voyager: Exploration, space and the third great age of discovery", Stephen J. Pyne, Penguin Books, 2010.
[33] Voyager, the interstellar mission, NASA JPL, California Institute of Technology, http://voyager.jpl.nasa.gov/, accessed 15 March 2014.
[34] Voyager the Interstellar Mission, Weekly report, 7 February 2014, http://voyager.jpl.nasa.gov/mission/weekly-reports/, accessed 15 March 2014.
[35] Tide, Wikipedia, 10 March 2014, http://en.wikipedia.org/wiki/Tide, accessed 16 March 2014.
[36] Sun, Wikipedia, 4 March 2014, http://en.wikipedia.org/wiki/Sun, accessed 16 March 2014.
[37] Moon, Wikipedia, 9 March 2014, http://en.wikipedia.org/wiki/Moon, accessed 16 March 2014.
[38] Tidal influence, Hyperphysics, C.R. Nave, Georgia State University, http://hyperphysics.phy-astr.gsu.edu/hbase/tide.html, accessed 16 March 2014.
[39] Friction, Wikipedia, 23 February 2014, http://en.wikipedia.org/wiki/Friction, accessed 16March 2014.
[40] Friction and coefficients of friction, Engineering toolbox, http://www.engineeringtoolbox.com/friction-coefficients-d_778.html, accessed 16 March 2014.
[41] Emergency brake (train), Wikipedia, 17 November 2013,

http://en.wikipedia.org/wiki/Emergency_brake_(train), accessed 16 March 2014.
[42] Typical stopping distances, Direct.gov.uk, http://www.direct.gov.uk/prod_consum_dg/groups/dg_digitalassets/@dg/@en/@motor/documents/digitalasset/dg_188029.pdf, accessed 16 March 2014.
[43] Traffic flow, Wikipedia, 10 March 2014, http://en.wikipedia.org/wiki/Traffic_flow, 16 March 2014.
[44] Ice skate, Wikipedia, 16 March 2014, http://en.wikipedia.org/wiki/Ice_skate, accessed 16 March 2014.
[45] Moment of inertia, Wikipedia, 15 March 2014, http://en.wikipedia.org/wiki/Moment_of_inertia, accessed 17 March 2014.
[46] Lists of moments of inertia, Wikipedia, 13 March 2014, http://en.wikipedia.org/wiki/List_of_moments_of_inertia, accessed 17 March 2014.
[47] Glossary of figure skating terms, Wikipedia, 26 February 2014, http://en.wikipedia.org/wiki/Glossary_of_figure_skating_terms, accessed 17 March 2014.
[48] home experiment: spinning box, Blogspot, 2 July 2008, http://skepticsplay.blogspot.co.uk/2008/07/home-experiment-spinning-box.html, accessed 17 March 2014.
[49] Rotational Stability, R. Fitzpatrick, University of Texas Teaching Notes, 31 March 2011, http://farside.ph.utexas.edu/teaching/336k/Newtonhtml/node71.html, accessed 17 March 2014.
[50] Helicopter, Wikipedia, 12 March 2014, http://en.wikipedia.org/wiki/Helicopter, accessed 17 March 2014.
[51] Bernoulli's principle, Wikipedia, 16 March 2014, http://en.wikipedia.org/wiki/Bernoulli's_principle, accessed 17 March 2014.
[52] Gyroscopes, Wikipedia, 10 March 2014, http://en.wikipedia.org/wiki/Gyroscope, accessed 17 March 2014.
[53] Precession, Wikipedia, 8 March 2014, http://en.wikipedia.org/wiki/Precession, accessed 17 March 2014.
[54] Milankovitch cycles, Wikipedia, 24 February, 2014, http://en.wikipedia.org/wiki/Milankovitch_cycles, accessed 17 March

2014.
[55] Axial tilt, Wikipedia, 16 March 2014, http://en.wikipedia.org/wiki/Axial_tilt, accessed 17 March 2014.
[56] Tachometer, Wikipedia, 17 March 2014, http://en.wikipedia.org/wiki/Tachometer, a7 March 2014.
[57] Sato, Y. & R. Packard, 2012: Superfluid helium interferometers, Physics Today, 65(10), 31 (2012); doi: 10.1063/PT.3.1749, http://159.226.72.19/share/reference/QuantumM/Sato-Packard12.pdf, accessed 17 March 2014.
[58] Mohs scale of mineral hardness, Wikipedia, 11 March 2014, http://en.wikipedia.org/wiki/Mohs_scale_of_mineral_hardness, accessed 17 March 2014.
[59] Hardness, Wikipedia, 15 March 2014, http://en.wikipedia.org/wiki/Hardness, accessed 17 March 2014.
[60] Hagen-Poiseuille Equation, Wikipedia, 7 February 2014, http://en.wikipedia.org/wiki/Hagen%E2%80%93Poiseuille_equation, accessed 17 March 2014.
[61] Viscosity, Wikipedia, 16 March 2014, http://en.wikipedia.org/wiki/Viscosity, accessed 17 March 2014.
[62] Surface tension, Wikipedia, 11 March 2014, http://en.wikipedia.org/wiki/Surface_tension, accessed 17 March 2014.
[63] Exploding water, a moment of science, Youtube, 18 June 2013, http://www.youtube.com/watch?v=8XaqCvOpKxs, 17 March 2014.
[64] Mass and volume of a housefly, Physics Forums, 25 October 2005, http://www.physicsforums.com/showthread.php?t=96681, accessed 17 March 2014.

Measuring the World, by John Austin

10. Pressure

Pressure = the force per unit area acting on a surface

10.1 Introduction

Force and pressure are often confused by the public, but pressure is actually the force per unit area. Perhaps the best example of this is atmospheric pressure, which, after the introduction of pressure units (§ 10.2) is described in § 10.3. In the case of the global atmosphere, the total force is very large (§ 10.4 - 10.5) but the pressure is a much more manageable 100 kN m^{-2} ($\approx$ 1 kg force cm^{-2}) or about 14.7 lb per in^2[1]. Atmospheric pressure falls off exponentially with altitude, so we are only comfortable within a few kilometres altitude of sea level. In aviation, the relationship between pressure and altitude is used to determine altitude[2]. It is then more normally referred to as the 'flight level', the nominal height above sea level which is distinguished from the true height above the ground. Hence, aircraft actually fly on constant pressure surfaces. As long as everyone uses the same system mid-air collisions between aircraft will remain rare.

Atmospheric pressure gives information about weather systems present, which accounts for its general importance and interest. Although many atmospheric concepts apply to other worlds, only the Earth is of specific interest here. Nonetheless, while not all astronomical bodies have atmospheres, they all have gravity, since gravity occurs because of mass (§ 9.5). The moon is perhaps the best-known example. This is often a point of confusion by some people, but it has weak gravity and almost no atmosphere. The escape velocity of most gases exceed that of the moon, so even if it had an atmosphere originally, it has long since escaped[3]. Note that Earth's atmospheric density is quoted in [3] as 1×10^{19} molec cm^{-3}, but the actual density is approximately Loschmidt's number[4], 2.687×10^{19}, almost three times larger, and well-known to atmospheric chemists like me. However, 1×10^{19} is from a NASA website, which should know better. Either way the moon's atmosphere is about 10^{-14} Earth atmospheres – zero for most purposes.

One howler which interestingly unites concepts in atmospheric pressure and the flow out of a bathtub is the rotation of the Earth. Should we expect to see a specific direction of rotation when we remove the plug of our bath tub? Conventional thinking would have us believe so, but a more balanced discussion occurs in § 10.6. Water pressure also increases rapidly with depth, so deep water diving has considerable risks associated with it. Water pressures and subterranean pressures are much higher than atmospheric and these bring their own particular problems and concerns (§ 10.7).

Sound waves (§ 10.8 – 10.10) are pressure waves in air, with loudness increasing with the magnitude of the pressure variations, while Earthquakes (§ 10.11) are pressure waves in the ground. Simple pressure concepts can be applied to motoring issues such as tyre pressures and engine exhaust pressures (§ 10.12 – 10.13). With relatively little effort, for example, proper tyre pressure can be calculated for a given load. The points are advanced, not as a comprehensive understanding of pressure in all its guises, but rather as an indication of how working entirely in the metric system makes it easy to tackle a diversity of problems with little effort. By comparison, § 10.14 takes a light hearted view of the effort involved in working with fps units.

10.2 Units of pressure

In the fps system of units, pressure has units of pounds per square inch (abbreviation psi or lb in^{-2} and there are other variants) commonly, even though the pound isn't a force, and the inch is not the standard fps length unit! To convert to a real force in poundals, you need to multiply by g (= 32.2 ft s^{-2}) but this is a less common fps unit. As noted above, atmospheric pressure is about 15 psi and typical tyre pressures are about twice as much. For the atmosphere, it is traditional to quote inches of a column of a biohazard (mercury), at least in the USA.

In the cgs system of units, pressures are measured in dyn cm^{-2} or in the SI system N m^{-2}. These are such important units that they have their own specific names. The dyn cm^{-2} is known as a barye[5], although I only discovered this while researching for this book. A more common cgs-like unit is the torr or mm of a mercury column. This of course depends on a biohazard, so is no longer politically correct. The N m^{-2} is also known as the pascal, Pa, after Blaise Pascal, the French physicist. Metric conversions are 1 dyn cm^{-2} = 10^{-5}N x 10^4 m^{-2} = 0.1 N m^{-2}. A common unit of pressure (cgs based) is the bar = 10^6 dyn cm^{-2}. To first order, atmospheric pressure is 1 bar, or 1000 mbar = 100 kPa. It is actually common to use hPa, even though this isn't a recommended SI prefix, since it allows the pressures to have the same numerical value as the no-longer favoured mbar or mb as it is sometimes written. One atmosphere pressure = 1013.25 hPa = 76 cm of mercury = 760 torr = 29.92126" approximately.

10.3 Atmospheric Pressure

The barometer was invented in the 17th century[6], and it was recognised that atmospheric pressure has a significant impact on weather. High pressures generally coincide with settled weather with clear skies (warm in summer, but it could be very cold in winter). Low pressures coincide with wet or stormy weather, and in the late summer, tropical storms are tracked by forecast services worldwide[7,8], because of their potential to inflict severe damage. In principle, you might expect air to flow from high pressure to low pressure and quickly equilibrate pressures. However, meteorological systems are very large and strongly influenced by the Earth's rotation. This leads, to a first approximation, to geostrophic flow[9], flow along the isobars (the contours of equal pressure on a meteorological chart). The actual wind has a small component from high to low pressure. Weather forecasts are now produced routinely by computer[10] applying the equations of physics of a fluid on the rotating Earth[9]. The methods used to solve the equations are extremely sophisticated and beyond the scope of this work. Forecasts are produced as 'weather maps' which include pressure at each point as a set of isobars. For the maps to show the meteorological systems, rather than a map of the altitude of the land, pressures need first to be converted to mean sea level.

The conversion is in principle straightforward and equals the pressure of a column of air which is $\rho g h$, where ρ is air density, g is Earth's gravity and h is the height of the station. For h = 10 m and the outside temperature is 0°C, then the correction is 1.293 x 9.81 x 10 = 126.8 N m^{-2}, which is 1.268 hPa for every 10 m. In practice we need to make a correction to the air density if the temperature is different, but this is a small correction. The air density is inversely proportional to the temperature in kelvins, so for example, if the outside air is 20 °C, then the air density is 1.293 x 273.15/293.15 = 1.204 kg m^{-3} so the pressure correction is 1.18 hPa for every 10 m of altitude difference. For example, if a station observes a pressure of 1000 hPa and is h = 100 m above mean sea level, then the pressure correction is *1.18 x 10 = 11.8 hPa, using an* air temperature of 20 °C. Except on stormy days, this is a large fraction of the pressure difference usually experienced across, say the UK, so it means that the altitude correction is crucial if we are to avoid a map instead of the hills and valleys of the country. In practice the pressure correction is not exact, since we are extrapolating air temperatures through solid Earth, which is not appropriate. In modern forecast systems, such as the UK Meteorological Office and other major forecasting centres, the true pressure is recorded together with the station altitude and the computer forecast model uses a terrain following coordinate system[11], known as sigma levels, so that pressure corrections do not need to be made. Instead, the pressure correction is applied after the computer forecast is produced to obtain a standard surface pressure map for interpretation by the forecaster.

Approximate atmospheric pressure as a function of height

Height (above sea level)	Pressure (hPa)	Height (above sea level)	Pressure (hPa)
0 km	1013.25	3 km	701
6 km	472	9 km	308
12 km	194	15 km	121

Mean sea level pressure is 1013.25 hPa and local values usually vary between 950 and 1040 hPa. Very strong storm systems can have central pressures below 950 hPa. Such storms are forecast very carefully, as they cause a huge amount of damage if they pass over land.

Atmospheric pressure decreases with height, as noted in § 10.1, because the atmosphere compresses under gravity. The effects can be quite large. The table above shows the approximate pressures corresponding to different altitudes. The decrease of pressure with height is approximately exponential: to a good

Measuring the World, by John Austin

accuracy $p = p_o e^{-z/H}$, where p_o = surface pressure, z is height and H is the scale height of the atmosphere = 7 km approximately. e is the base for natural logarithms = 2.718..... The relationship between pressure and height is easily derived using the integral calculus, if you assume an isothermal (constant temperature) atmosphere: for an infinitesimal column of height dz, the change in pressure is $dp = - \rho g\ dz$. So it is just a matter of integrating this from the surface to a height z, recognising that ρ itself is proportional to pressure for constant temperature[12]. In practice, the integration of the equation is done numerically assuming the height varying value of the temperature.

Commercial jet aircraft typically fly at 9 km altitude (300 hPa), or higher but the cabin needs to be pressurised for comfort to passengers. Internal pressure is usually set at about the equivalent of 2 km altitude[13]. Nine km is also close to the highest mountains on Earth (e.g. Everest at 8829 m). Since the pressure is barely 30% of that at sea level, special training is required to complete the ascent without oxygen[14]. Because of the strong dependence of pressure on height, the data in the above table can be used to provide approximate altitudes. In practice, commercial aircraft cruise on specified pressure surfaces (given by aircraft control, AC). In preparation for landing, AC provide pilots with the surface pressure (QNH)[15], so that an accurate altitude can be determined.

Even at an altitude of 80 km, the atmospheric pressure is still 0.01 hPa. Although this is very small there are still constituents present which contribute to a vast array of important processes and features, including noctilucent clouds[16], in the altitude range 76 to 85 km. Other interesting phenomena, such as the aurora[17] occur still higher in the atmosphere. An altitude of 100 km is often nicknamed 'the edge of space'[18], and more precisely referred to as the Karman line. It seems that astronauts are given their wings for ascending to the edge of space, but different US organisations cannot agree whether it should be above 50 miles or higher still. One can of course understand pilots or other worthies being given their 'wings', but astronauts? The whole point is that wings are ineffective in space. Some have even been given there 'astronaut wings' posthumously. I just find it odd that people take such things so seriously.

Atmospheric pressures can be compared with the 'vacuum cleaner': something of an exaggeration as the internal pressure is not much lower than 0.9 atmospheres, with only about 100 hPa reduction[19]. By contrast, the hardest laboratory vacuum produced[20] was 10^{-13} Pa ≈ 10^{-18} atmospheres. Even at these pressures, the sample still has 30,000,000 molecules of mixed gases per m^3. Deep space has just a few molecules m^{-3}[21].

10.4 Measuring Atmospheric Pressure

Historically, atmospheric pressure was measured using an evacuated tube placed in a bowl of mercury[22]. Atmospheric pressure pushes down on the open surface of mercury supporting the column of mercury which rises in the tube. Atmospheric pressure, p, is then given by $p = \rho g h$, where ρ is the density of mercury, g is the acceleration due to gravity and h is the height of the column. Astonishingly, in the USA, the pressure is still quoted this way, but the conversion of the height of a real mercury column to pressure units requires a number of factors and adjustments. A mercury barometer is a lovely apparatus that deserves veneration as much as an antique clock. Measuring the column length is the first step, and the use of a 'vernier' scale[23] improves the precision to 0.1 mm or 0.01 in. The vernier was invented by Pierre Vernier in the 1600s. It consists of a short scale moveable alongside the main scale. The short scale has divisions 9/10 of the size of the smallest division of the main scale. This could be 1 mm, with vernier divisions 0.9 mm apart. The mercury column has a scale next to it to allow its length to be determined. The zero of the vernier is aligned with the top of the meniscus of the mercury column, and the number of divisions of the vernier are counted up to the point at which the vernier and the main scale coincide. This number then indicates the additional length in units of 0.1 mm. For example, suppose the actual length is 75.83 cm. Using the main scale only would give 75.8 cm, but the actual column would extrude 0.3 mm above the meniscus. If the zero of the vernier is put at the top of the meniscus, the graduations of the vernier would not coincide at the zero point. The next points on the vernier would be at 75.92, 76.01 and 76.10. So the third division above the vernier zero would coincide with a graduation on the main scale, and the length is 75.8 cm from the main scale plus 3x0.1 mm from the vernier. Some verniers have been designed with even higher precision in mind, with for example vernier divisions of 19/20 of the size of the main divisions.

Once the mercury column has been measured, it is necessary to make three further corrections[24]. The first is for the saturated vapour pressure of mercury. The space in the tube above the mercury column actually exerts a downward pressure on the column due to the presence of mercury vapour. This results in an upward correction of the column measurement, an amount which is dependent on the temperature of the mercury. Often, barometers are supplied with a crude thermometer to make this correction, which is about 2 cm. The second correction is for the thermal expansion of mercury. This lowers the density so the correct density for the temperature already measured for the mercury needs to be used in the pressure calculation. The local pressure can now be calculated using the previous equation $p = \rho g h$, where ρ is the correct density and h is the measured column plus the correction for the mercury vapour pressure. Of course the local value of g should be used rather than some averaged value. If the pressure is to be kept in mercury column lengths then the pressure can be adjusted to the appropriate column with standard gravity and density of mercury at 0 °C. For atmospheric purposes, a third correction is needed: to reduce the column to the equivalent value at sea level. This additional correction is $\Delta p = \rho g h$ where now ρ is the density of air and h is the altitude of the observing station. Note that the air density needs to be computed at the temperature of the outside air, not the temperature of the mercury, or essentially room temperature. The correction works out at about 1.3 % for every 100 m altitude. The calculation of atmospheric pressure this way is quite an ordeal, even ignoring the fact that mercury is a biohazard!

Mercury barometers have now been phased out by the UK Meteorological Office[24] and the current practice is to use a 'precision aneroid' barometer. In an aneroid barometer, an evacuated chamber is compressed by the atmosphere, and the movement of the edge of the chamber is measured. Typically this might only be 1 mm for a 100 hPa change in pressure, so the movement is magnified by a sequence of gears pivots and levers. As the pressure changes, a pointer fixed to the chamber moves along a calibrated scale. In a 'precision' aneroid, used for professional purposes, the response of the mechanical movement inside the barometer has been improved to be more responsive to atmospheric changes and to allow measurements to be made to a precision of 0.1 hPa.

10.5 The Mass of the Atmosphere

The mass of the atmosphere can be calculated very simply by multiplying the mean force per unit area (pressure) due to the atmosphere by the surface area. Dividing the force by the acceleration gives the equivalent mass $M = 4\pi R^2 p_o/g$, where p_o is the mean surface pressure and R is Earth's radius. Taking $R = 6.37 \times 10^6$ m, and $p_o = 995$ hPa, we obtain $M = 4\pi \times (6.37 \times 10^6)^2 \times 0.995 \times 10^5/9.81 \approx 5.17 \times 10^{18}$ kg. By human standards, this is a very

Measuring the World, by John Austin

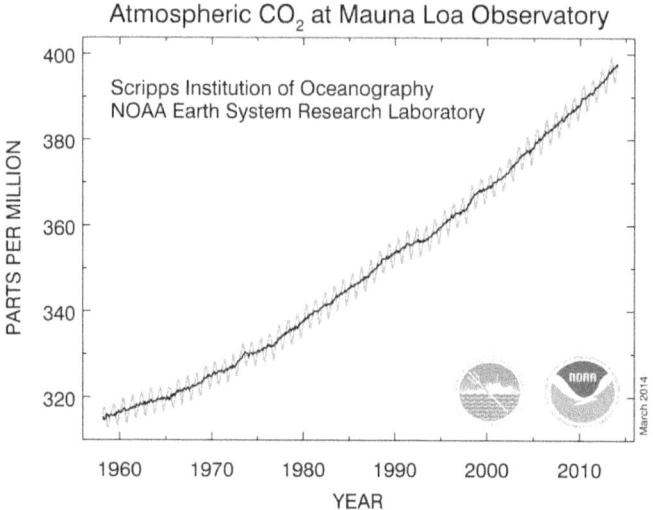

The increase of CO_2 in the atmosphere since the late 1950s[25]. Measurements have been made as a joint venture between Scripps IO and NOAA ESRL and have been shown up to February 2014. the monthly mean values at Mauna Loa observatory in Hawaii are shown in red, and the annually averaged values shown in black.

large mass, and it might be surmised that humans can't possibly influence the atmosphere on a global scale. Think again. For example, CO_2 was present in only small quantities, 270×10^{-6} (270 parts per million or 270 ppm) volume mixing ratio in the pre-industrial atmosphere. CO_2 mass mixing ratio is larger by a factor 44/29 so the atmospheric mass of CO_2 was $5.17 \times 10^{18} \times 270 \times 10^{-6} \times 44/29 = 2.12 \times 10^{15}$ kg. However, since industrialisation began, humans have already added 578 million tonnes of carbon into the atmosphere[26]. That may not sound too bad, but the figure quoted is tonnes of *carbon*, and the actual mass of CO_2 released would have been higher by a factor 44/12 or 2.12 trillion tonnes = 2.12×10^{15} kg, or coincidentally the same value as was present pre industrially. Half of the release has gone into the oceans, and the rest has stayed in the atmosphere and increased the concentration by about 50% to almost 400 ppm[25, see diagram]. This increase is having major impacts on the climate and will continue to do so for the rest of the 21st century, as reported at length in the Fifth Assessment Report of Intergovernmental Panel on climate Change[27]. Politicians have already agreed that a 'safe climate change' limit would be a 2 °C warming and the best scientific evidence suggests that this would be attained with total release of about 1 trillion tonnes of carbon[25]. This would require the current release eventually to slow to a trickle. Having already reached almost 60% of the target, it follows that with current inactivity by politicians we have no prospects of reaching the so called safe limit, and we will need to be prepared for the additional cost to our economy such as highlighted by the Stern report[28].

10.6 Earth's Rotation: winds and bathtubs

For large scale meteorological systems, the pressure gradient acceleration is balanced by the acceleration due to the rotation of the Earth[29]. The acceleration due to the rotation of the Earth is known as the Coriolis acceleration, named after the French atmospheric scientist. Setting these two accelerations equal, we find that

$$fu = \nabla p/\rho.$$

∇p is the pressure gradient, $f = 2\Omega \sin \vartheta$ for latitude ϑ, and $\Omega = 2\pi/1$ day^{-1} = the rotation rate of the Earth. In this case u is known as the geostrophic wind[9], which is a good approximation to the actual wind in middle and high latitudes, but the relationship breaks down near the equator. The wind direction in relation to the pressure system was enunciated by Buys Ballot in the law of that name: if you stand with your back to the wind, in the Northern hemisphere the low pressure is to your left[30]. In the Southern hemisphere, the low pressure is to the right. The relationship between pressure gradient and wind strength is commonly noted indirectly by weather forecasters working from a meteorological chart. The isobars are drawn with the same contour interval each time, usually 4 or 8 hPa, and for regions where 'there are a lot of isobars', this means the pressure gradients are high and so the winds are strong.

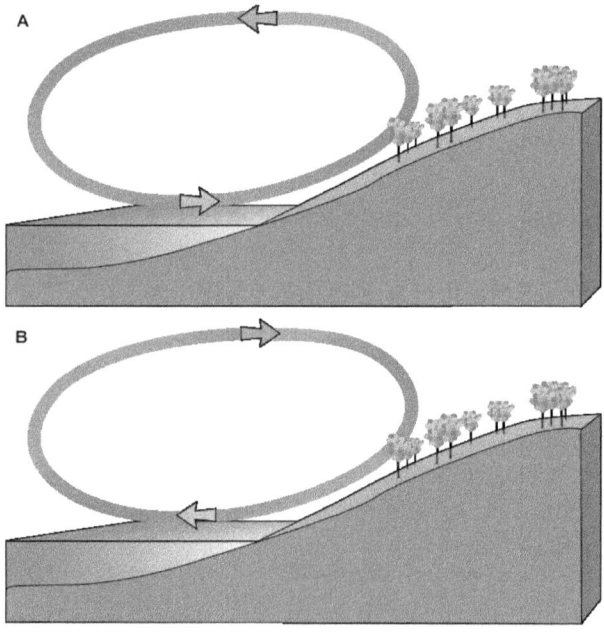

The formation of the sea (A) and land(B) breeze. The breezes are each driven by higher temperature which lead to lower pressure and a direct circulation from high to low pressure. Earth's rotation is negligible. Author Jesus Gomez Fernandez, 19 June, 2005[31].

Measuring the World, by John Austin

These processes work on very large scales – about 1000 km. On a much smaller scale, that of a beach, about 1 – 10 km, the Coriolis force is much smaller than the pressure gradient force. We then get effects such as the sea breeze (wind off the sea) during the day time, and land breeze (wind off the land) during the night[31]. These effects occur from direct rising motion over the warm area (land during the day, sea during the night) and a return circulation at the surface. The return circulation occurs directly from high to low pressure. What this indicates is that the Earth's rotation is only important for scales much larger than 10 km.

We turn now to rotating liquids in a sink or bathtub, when you pull the plug out. It is something of an urban myth that the direction of rotation is related to the Earth's rotation, with the water flowing counter-clockwise in the Northern Hemisphere, and clockwise in the southern Hemisphere. The behaviour of the sea breeze leads us to question this, even ignoring the other evidence[29]. Forty years ago, I remember seeing a television programme in which it was rigorously tested. The tank was circular, about 10 m in diameter. Water was added to a depth of a metre or so, in a clockwise direction and allowed to settle for several days.

When the plug was removed, the water left the tank without a preferred direction of rotation. The implication is that any rotation of the fluid is related to the initial rotation as well as the shape of the sink or bathtub, but under extremely careful conditions, a tank might acquire some of the Earth's rotation.

I had an amusing experience in 2001 when my wife and I visited Kenya. During part of our tourist activities, we passed a village that was on the Equator. Which was marked by a line on the ground. Presumably, it could easily have been incorrectly placed by 5 – 10 m. One of the locals was demonstrating the Coriolis force with a small bucket of water (diameter 25 cm or so). Once he had a reasonably-sized participating audience, he showed water being poured into the bucket while standing in the 'Northern Hemisphere'. He demonstrated the anti-clockwise rotation of the water on removing the plug from the bottom of the bucket. He then walked 20 m into the 'Southern Hemisphere' and repeated the experiment, when lo and behold the water left the bucket with a clockwise rotation. I 'proudly' received my certificate for 'crossing the Equator' but immediately started wondering how this had worked despite the failure of the experiment under carefully-controlled laboratory conditions. It was only later that I worked out how he managed to perform the trick: when he is in the 'Northern Hemisphere' he filled the bucket with his right hand; when he was in the 'Southern Hemisphere' he filled it with his left hand, thereby imparting the desired rotation in both cases. By doing this consistently he was assured of a steady income selling certificates to twits like me! In actual fact, the Coriolis force is so weak near the equator that it doesn't even work for 1000 km sized weather systems!

10.7 Water Pressure and Subterranean Pressure

The pressure, p, due to a column of fluid of height h and density ρ is $p = \rho g h$, where g is the acceleration due to gravity. Hence, the pressure increase per metre of water is ρg. Putting ρ = 1000 kg m^{-3} and g = 9.8 m s^{-2}, the pressure increase is $\approx 0.1 \times 10^5$ Pa $\approx$ 0.1 atm. The rapid increase of water pressure with depth makes pressure gauges an accurate measure of depth. The rapid increase arises from the much higher density of fresh (1000 kg m^{-3}) or sea water (1100 kg m^{-3}) compared with air (1.3 kg m^{-3}), a factor of over 800. Thus for a mere 10 *m* water depth, there is an additional atmosphere of pressure. When diving, the pressure of even quite shallow water needs to be considered.

The complications that arise due to the high pressures in deep sea diving are now well documented[32], although it took many decades to complete this research. Issues such as the breathing mixture, and depressurisation need to be carefully considered. At high depths, if air is breathed, nitrogen is absorbed into the blood. If the subsequent ascent is then too fast, the nitrogen is released as bubbles in the blood, producing the 'bends', a painful and possibly fatal condition. If ascent is more gradual, then the nitrogen can be exhaled from the lungs without bubbles forming in the blood. Safe ascent rates are now well established, and include the need for periodic rests. The problem cannot be easily solved by removing the nitrogen from the breathing mix since it would need to be replaced by another gas, and most breathable gases have a narcotic effect to a degree. Once the pressure reaches a certain point, narcosis would set in. The narcosis of a gas is believed to be related to its lipid solubility[34,35], as in the following table.

Gas	Solubility mL/mL	Gas	Solubility mL/mL
Xenon	1.90	Oxygen	0.11
Nitrous oxide	1.56	Nitrogen	0.052
Carbon dioxide	1.34	Hydrogen	0.042
Krypton	0.44	Helium	0.015
Argon	0.15		

In the above table, nitrous oxide is a well-known narcotic at atmospheric pressure while carbon dioxide is also known to be hazardous in an unventilated environment. The lowest five values in the table are much lower than the others, and are except for Argon are part of a plausible breathing mixture. Argon would not be considered because it is likely more narcotic than oxygen. Oxygen narcosis leads to dizziness, vomiting and convulsions so even compressed air with an oxygen concentration of only 20% cannot be used below about 50 m depth. Nitrogen narcosis is similar to drunkenness with an inability to reason and a lack of control. With compressed air, containing 80% nitrogen, this sets in earlier at about 30 m depth. The best solution to these problems is to substitute most of the nitrogen in the air by an inert gas such as helium (He), which has a lower solubility in the above table. At 1.0 - 1.6 atm partial pressure of O_2, oxygen narcosis starts to set in, so the O_2 in the He/O_2 breathing mix needs to be reduced accordingly at higher depths than about 60 m. Breathable mixtures are often a combination of $He/N_2/O_2$ and referred to as trimix[36], which are graded for different depths.

Some whales are capable of diving down to a depth of perhaps 3 km[37], although 400 m is more common, for what are perhaps some of the most remarkable adaptations in the animal world. Sperm whales for example, are well-adapted to swimming under water and usually swim for 35 minutes without surfacing but can go under for 90 minutes. They are thermally insulated with a thick layer of blubber, which also increases their buoyancy. Sperm whales can hold their breath for up to 2 hours. The whales have 41% oxygen supply in the muscles, compared with 13% for humans. The underwater duration of some species is also extended by their ability to lower heart rates and reduce the blood flow to inessential muscles. After a long dive, these whales will usually take several breaths to recharge their systems. Over the period of their lifetime, whales do sometimes show some signs of decompression problems but they are generally adapted to deep dives by their improved metabolism and physiology[37]. At still higher depths, creatures have been discovered which live deep in the ocean, several thousand metres below the surface[38]. Such wildlife has needed to adapt to the extreme pressure, absence of light and severely limited food resources.

Subterranean pressures are even higher than the equivalent sea depth because rock has a higher density than water. Extreme pressures and temperatures have conspired to create the internal Earth structure that is described elsewhere[39].

Measuring the World, by John Austin

10.8 Sound Waves

What we perceive as sound is actually a pressure wave in the air[40]. The larger the amplitude of the pressure wave, the louder is the sound, which is measured in bels = $2log_{10}(p/p_o)$ where p is the sound pressure, and p_o is a standard pressure = 2×10^{-5} Pa and 1×10^{-6} Pa in water. This scale reflects the tendency of our ears to be sensitive to the logarithm of the pressure wave amplitude, rather than proportional to the amplitude itself. Sounds are more usually quoted in decibels = 0.1 B, so a loud sound = 100 dB = 10 B, corresponding to p = 2 Pa, which is comparable to fluctuations in atmospheric pressure over periods of a few minutes.

The bel is named after Alexander Bell, the Scottish engineer. It is not strictly an SI unit, but together with napier (a logarithm to the base e) it is often used with SI units[41]. While the above definition applies to the pressure level of sound, bels are often used in the context of electrical signals and the base level for the comparison then needs to be specified. The softest audible sound is about 0 dB, hence the choice of the value of p_o, although sound perception is frequency dependent and varies with age and gender[42]. Normal conversation is about 60 dB while extremely loud music is about 120 dB. A jet engine during maximum thrust makes a noise of about 140 dB at close range which is at the threshold of pain. Other sound levels are whispering (20 dB), telephone (70 dB), vacuum cleaner (80 dB), and circular saw (100 dB). The apparent loudness of a sound depends on its frequency. For example, humans are insensitive to ultrasound (frequency > 20 kHz) and most sensitive near 4 kHz[43] and the 0 dB loudness limit is just detectable at 1 kHz. In the animal kingdom, the frequencies emitted and those audible vary, which provides an interesting discussion in itself, but this is beyond the scope of the book. As a short summary, the table below provides a short sample of emitted and audible frequencies (Hz) by assorted wildlife[44,45,46], although the values are somewhat approximate.

Sound Frequencies (Hz) in the animal kingdom

Animal	Emitted	Audible	Animal	Emitted	Audible
Human	85-11000	20 – 20,000	Bat	15,000-90,000* (typically)	15,000-90,000* (typically)
Dog	450-1100	40 – 60,000	Bird	2,000-13,000*	250-21,000 *
Cat	760-1520	55 – 79,000	Mouse	Up to 40,000	1,000-70,000
Dolphin	75 – 150,000	75-150,000	Grasshopper	7,000-100,000	100-100,000

* Depending on species

In 2001, my wife and I went swimming with dolphins during a memorable experience in Kaikoura, New Zealand. We dressed in a wet suit early in the morning, around sunrise, and after some quick instruction, we boarded a small boat. We were taken out into the surrounding bay and put into the deep ocean water some distance from the dolphins. The idea was that by our making squeaking noises and splashing around, the dolphins would come to investigate . The squeaks were at the lower end of their hearing range and it worked very well. To have dolphins swim within touching distance was a memorable experience.

There is a clear relationship between people and their dogs and cats. The animals can hear human sounds and make themselves heard. In addition, they hear in the ultrasound range so can respond to whistles at these frequencies, although cats are not known for obeying their owners!

The loudness of sound can seem different at different

frequencies, even though it may be the same loudness measured in dB. Accordingly, the phon is defined as the intensity in dB of a 1 kHz tone that *seems* to be equally loud[47]. Clearly this is a subjective measure, but tests have been made for the average person without hearing problems and a hearing response curve has been agreed. For example, 90 dB @ 20 Hz ≈ 20 dB @ 1000 Hz = 20 phon, on the extreme left of the curve.

Pressure waves are solutions to the equations of the atmosphere, as are meteorological waves and systems. Sound waves would dominate the solutions and so are routinely filtered from weather forecasts[48]. The usual procedure is to change the atmospheric equations to use pressure as a coordinate instead of height. The relation between pressure and height is then $\Delta p = -\rho g \Delta z$ where Δp is the change in pressure due to a small vertical height change Δz and ρ is the air density. g is the acceleration due to gravity. This is known as the 'hydrostatic approximation'. Recently, forecast models have been designed with improved accuracy using height as a vertical coordinate. The hydrostatic approximation is no longer assumed, so sound waves need to be filtered from the forecast using a different method[49].

10.9 The speed of Sound

Theory predicts that the speed of sound in a gas is $(\gamma RT)^{\frac{1}{2}}$[50]. R is the gas constant for the specific gas and in SI units is measured in J kg^{-1}K^{-1}. T is temperature measured in kelvins and γ is a constant which depends on the molecular structure of the gas. For most gases, $\gamma = 1.4$ to a very good approximation. For the 'noble' gases such as helium, argon etc., $\gamma = 5/3$. For air, only traces of noble gases are present, $R = 287.06$ J kg^{-1}K^{-1} and $\gamma = 1.4$, with measured values are in the range 1.3991 to 1.403. Hence the speed of sound in air at room temperature ($T = 293.15$ K, or 20 °C) is 343.2 m s^{-1} (assuming $\gamma = 1.400$). At the altitude of commercial flight, the atmospheric temperature is sometimes as low as 200 K, giving a much lower speed of sound of 283.5 m s^{-1}.

Aircraft speeds are frequently measured relative to the speed of sound in air (Mach number, after the Austrian physicist Ernst Mach)[51]. As the aircraft speed approaches Mach 1, air is compressed ahead of the aircraft producing a pressure wave generated by the aircraft. Exceeding the sound barrier requires a significant expenditure of energy and a different wing design to allow proper aircraft handling. The pressure wave generated can causes damage to buildings on the ground, so is generally avoided near cities. A meteorite entered the Earth's atmosphere spectacularly over Chelyabinsk, Russia on 15 February 2013[52]. Thanks to the common use of video cameras on people's cars, the phenomenon was extremely well captured by a host of people. The videos show the curved path of the meteor burning brightly as it passed through the outer reaches of the atmosphere, and as it approached the ground, windows in buildings were smashed due to the sonic booms that resulted. With footage like this, one can understand how some people might believe in little green men from outer space!

The speed of sound in liquid and solid material is much higher than in air, and depends on the density and compressibility of the material, according to the theory summarised in [50]. The speed of sound in fresh water is dependent on temperature and is 1497 m s^{-1} at 25 °C. In sea water the speed of sound is dependent on temperature and salinity and is typically about 1560 m s^{-1}. Sound waves in solids are related to earthquakes and are described in 10.11.

10.10 The Doppler Effect

The Doppler effect is the well-known effect with sound, in which a fast moving object such as a car appears to increase in pitch as it approaches, and decrease in pitch as it recedes[53]. The simple explanation is that from the perspective of the observer, the wavelength of the sound is being shortened on approach and because of the fixed speed of sound, the frequency needs to increase to compensate. The reverse happens when the vehicle recedes: the wavelength of sound increases, and the frequency or pitch decreases. The audible frequency, f', is given by $f' = f(1 + v/c)$, where f is the original frequency, v is the speed of the car or moving object and c is the speed of sound. In this expression, when the car is approaching v is taken to be positive, and when it is receding the speed is taken as -v. For sound waves on a warm day, c = 343 m s^{-1} and if v = 25 m s^{-1} (56 *mph*) then a frequency of f = 256 Hz (say, middle C depending on the precise scale) will be shifted to f' = 274.7 Hz, distinguishable audibly from the original frequency.

Similar processes occur with electromagnetic (EM) waves. For example, in police speed traps, the Doppler effect on a radar beam is used to determine vehicle speeds. A radar wave is directed at the vehicle and by measuring the frequency of the radar reflection the speed can be determined. The reflected frequency, f', is given by a slightly more complicated relationship[54]

$$f' = f(1 + v/c)/(1 - v/c)$$

where now c is the speed of light (the same as the speed of radar). When a radar beam is being used, the frequency change is very small, $f - f' \approx 2fv/c$, but can be measured directly by detecting the 'beat frequency', $f - f'$.

10.11 Sound Waves in Solids: Earthquakes

Unlike gases and liquids, there are two types of sound waves in solids, compression and shear waves, and each of these has a different velocity[50]. The compression waves, also known as

Measuring the World, by John Austin

P-waves or longitudinal waves have higher values than the shear waves, also known as S-waves or transverse waves. Some typical velocities in m s^{-1} are [55,56,57] as follows:

Solid	P-wave speed	S-wave speed	Solid	P-wave speed	S-wave speed
Aluminium	6420	3040	Diamond	-	11650
Iron	5950	3240	Brick	4176	-
Copper	4760	2325	Concrete	3200-3600	-
Silver	3650	1610	Glass	3962	-
Tin	3320	1670	Rubber	40-150	-
Lead	2160	700	Wood	3300 - 3960	-
Stainless Steel	5790	3235			

Earthquakes are essentially sound waves in the Earth's crust. The P-waves which have velocities of 6.8 km s^{-1}, and the S-waves have velocities of 3.8 km s^{-1}. Rayleigh and Love waves are confined to the surface. At depths of 1000 km, wave speeds increase by a factor of 1.5 or more, since the velocities increases as the square root of temperature. The increase from 300 K (27 °C) near the surface to 600 K (327 °C) deeper in the crust would give rise to an increase of a factor of √2 = 1.41 in the increase in velocity. At still higher temperatures in the Earth's mantle, the P-waves travel at about 13 km s^{-1}[58]. In earthquake-prone countries such as Japan, specially resistant buildings need to be constructed. This includes the use of synthetic materials to absorb the sideways motion induced by the earthquake.

Earthquakes in the ocean bed can lead to deadly

Measuring the World, by John Austin

tsunamis such as the one which reached land at the Fukishima nuclear power station in Japan on 11 April 2011[59]. The deadliness of tsunamis comes from their hidden force. Although the ocean floor is displaced by only a few centimetres, several kilometres column depth of water may be affected. The column propagates, and as the water depth decreases close to shore, the wave magnifies.

Seismographs are used to locate and measure earthquakes[58], essentially using triangulation from a global network of instruments and timing the arrival of the waves. Details of the Earth's interior can also be deduced using information from seismic waves. Not all earthquakes are natural: seismographs are used to detect the earth tremors associated with test nuclear explosions. Also, one of the environmental concerns about fracking (extraction of gas by forcing steam into sedimentary rocks) is that the procedure probably causes micro-earthquakes which could cause cumulative damage[60], and seismographs can be used to monitor this. The strength of an earthquake is measured on three scales: the Moment scale, the Richter scale, and the Mercalli intensity scale. The moment scale is somewhat complicated[61] and was designed to solve the problems with the use of the Richter scale. However, the latter is more often quoted by the media. The Richter scale measures crustal movement and is based on a log scale $R = log_{10}(A/A_o)$, where A is the amplitude of the seismic wave and A_o is a scale size. The amplitude needs to be determined by a group of seismographs to determine the epicentre of the earthquake. With Richter's original scale, the purpose was to ensure that all values should be greater than zero and A_o was chosen to be 1 μm at a distance 100 km from the epicentre. Note that the scale is a logarithmic function so that an increase of 1 in the Richter scale would require a factor of 10 increase in amplitude. A 4 PJ nuclear detonation would yield an effect equivalent to 7.2 on the Richter scale[61] and smaller yields are easily detectable.

The Mercalli intensity scale[62] is a qualitative scale describing the local effects of an earthquake, varying from a barely felt intensity (level I) to total damage (level XII). As if to emphasise the lack of a mathematical basis, the levels are indicated by Roman numerals (after all, the Romans must have struggled with basic arithmetic with a system like theirs).

10.12 Tyre Pressures

A question of every day practical use is: What tyre pressure is needed to support a car of known weight? If the area of tyre in contact with the road is A then the weight supported is pA, where p is the tyre pressure, measured above ambient. The typical tyre 'footprint' is an area width of a tyre is 15 cm, by about 10 cm, the length in contact with the road for each tyre. Hence, the total area of tyre supporting the car is $A = 4 \times 0.15 \times 0.1 = 0.06$ m². Hence the weight supported = mass x g. For a 1 tonne loaded vehicle, mass = 1000 kg, hence 1000 x 9.8 = pA or, p = 9800/0.06 ≈ 1.6×10^5 N m^{-2} approximately. Atmospheric pressure is about 1 x 10^5 N m^{-2}, so the tyre pressure (above ambient) needs to be 1.6 atmospheres. Usual recommended tyre pressures are about 2.2 atm above ambient pressure, so the corresponding area of contact would be slightly lower than in the example above, at 0.044 m². Actual tyre footprint details are difficult to obtain, but for a US passenger car, the tyres are about the same size in total, but inflated to a higher pressure [63]. Suppose now, though that you decided to drive your mother-in-law, spouse and an additional adult. Then the extra load would be about 200 kg, say. If you kept the tyre pressures the same, then to carry the load the tyres would spread out from 0.044 to 0.053 m² as the total mass has increased from 1.0 to 1.2 t. Alternately, you could add a bit more air to the tyres so that the pressure excess above ambient is increased by

20% to keep the contact area the same. This requires a pressure increase from 32 to 38 lb in^{-2} or from 2.2 to 2.6 kg cm^{-2}. Unfortunately, most of the pressure gauges I've seen are marked in lb in^{-2} and kg cm^{-2} but I suppose there might be some marked in proper pressure units such as bars. In this case, 1 kg cm^{-2} is very close to 1 bar (2% error).

An obvious point, not publicised much, is if you run on slightly soft tyres, the road contact area increases and hence braking is faster on a dry road. On a wet red, it might be anticipated that soft tyres are less effective because the tread becomes more compressed increasing the likelihood of aquaplaning. Soft tyres would also lead to increased damage and wear to the tyres. Putting too much pressure in tyres lowers contact area, decreasing friction and improving energy use (miles per gallon) but at the cost of safety (longer braking times on a dry road).

10.13 Engine Exhaust Pressure

Continuing with the vehicle theme, the pressure and temperature of car engine exhaust needs to be carefully controlled Engine exhaust is initially at high temperature and if released directly into the atmosphere would expand quickly producing a loud, sharp noise. The noise is reduced by the silencer or muffler in US-speak[64] which has a sequence of tubes along which the exhaust passes. The exhaust pressure is reduced gradually, and its heat is released to the silencer before the exhaust is released to the atmosphere. The silencer also has a chemical catalyst (usually platinum and other materials) on its inside surfaces to remove unburnt hydrocarbons, carbon monoxide and nitrogen oxide from the exhaust before release.

Measuring the World, by John Austin

10.14 Pressure Calculations with FPS Units

To end on a light note, could the above calculations be done with the fps units? Well, of course they can, but read on.

Barometer corrections: If the height of the station is 1000 ft and the barometer reading is 29.50 in and the mercury in the barometer is at 20 °C, how do you do it? Its a bit of a mess. You have to know the density of air in lb ft^{-3} so that you can do the altitude correction. You also have to calculate the air density at the air temperature to the proper density at the observed Rankine temperature. Rankine temperature is so obscure, I haven't even mentioned it. Then you need to know the density of mercury at 20 °C to convert to the 0 °C standard density. Then you need to know the saturated vapour pressure of mercury in inches of mercury. You might need to make a correction for local gravity compare with the 32.17 ft s^{-2} standard. Good luck!

Mass of the atmosphere: In one way this is easier than in metric, as the pressure of the atmosphere is given in mass units not force units so the mass is the area x pressure. However, this one advantage is lost by having to convert the radius of the Earth from miles to inches and then dividing the resulting mass in pounds by 2240 to get to tons. I still wouldn't want to do it, as there are too many conversion factors to keep track of, and it's just far easier to do the whole calculation in metric and convert to fps at the end. It's best to stay in metric.

Calculation of the amount of man-made CO_2: this has the same sort of problem as the calculation of the mass of the atmosphere. There just seem to be too many conversion factors to keep track of.

Calculation of water pressure for diving: This is probably just as easy in fps as metric. But do you know the density of water? I don't without looking it up, whereas everybody knows that water is 1 g cm^{-3} or for the more modern person 1000 kg m^{-3}

Measuring the World, by John Austin

(allow a bit more for sea water). The other thing is that intoxication limits for gases seem to be given in bars of pressure, so are you flexible enough to convert to your beloved psi? This experience could be better than taking cannabis, if it doesn't kill you.

Speed of sound in air: Where do you find the gas constant in funny units? If you find some obscure engineering source, can you be confident it is right, and if so do you know enough about Rankine temperatures?

Speed of sound in water: If you're trying a bit of echo location to find that long lost undersea treasure, you can work in ft s^{-1} for the speed of sound, but you will need your calculator to divide the distance found by that mysterious number 5280.

Calculation of tyre pressures: The load calculation is the same as in metric as you just do the calculation pressure x area. Pressure will be in lb in^{-2} and the tyre footprint will be in square inches so you will get a weight in pounds. You might then have to divide by various funny numbers like 112 or 2240 to get it into a more useful loading. The real problem comes when you want to change the load and people's weights are reported to you in stones, possibly one of the silliest unit around. At least US citizens quote their weight in pounds and save themselves a bit of aggravation. Can you still remember how many pounds are in a stone? Is it 13, or is it 14? No that doesn't sound right. Perhaps it's 15 or even 16. No, that's not right either: that's the number of ounces in a pound. Or are there 14 ounces in a pound? You tell me! [See Chapter 7!]

References

[1] Atmospheric pressure, Wikipedia, 6 March 2014, http://en.wikipedia.org/wiki/Atmospheric_pressure, 18 March 2014.
[2] Flight level, Wikipedia, 25 February 2014, http://en.wikipedia.org/wiki/Flight_level, accessed 18 March 2014.
[3] Atmosphere of the moon, Wikipedia, 29 January 2014,

Measuring the World, by John Austin

http://en.wikipedia.org/wiki/Atmosphere_of_the_Moon, accessed 18 March 2014.
[4]Loschmidt constant, Wikipedia, 26 February 2013, http://en.wikipedia.org/wiki/Loschmidt_constant, 18 March 2014.
[5] Centimetre-gram-second system of nits, Wikipedia, 9 March 2014, http://en.wikipedia.org/wiki/Centimetre-gram-second_system_of_units, accessed 18 March 2014.
[6] History of the barometer, Barometer Fair, http://www.barometerfair.com/history_of_the_barometer.html, accessed 18 March 2014.
[7] About the National Hurricane Center, US National weather Service, National Hurricane Center, 26 August 2013, http://www.nhc.noaa.gov/aboutintro.shtml, accessed 18 March 2014.
[8] Tropical Cyclone warning services, Australian Government Bureau of Meteorology, http://www.bom.gov.au/cyclone/about/warnings/, accessed 18 March 2014.
[9] Geostrophic wind, Wikipedia, 21 October 2013, http://en.wikipedia.org/wiki/Geostrophic_wind, accessed 18 March 2014.
[10] Forecasting by computer, European Centre for Medium Range Weather Forecasts, 22 January 2013, http://www.ecmwf.int/about/overview/fc_by_computer.html, accessed 18 March 2014.
[11] Numerical weather prediction, Wikipedia, 3 February 2014, http://en.wikipedia.org/wiki/Numerical_weather_prediction, 18 March 2014.
[12] Vertical pressure variation, Wikipedia, 29 January 2014, http://en.wikipedia.org/wiki/Vertical_pressure_variation, accessed 19 March 2014.
[13] International travel and health, World Health Organisation, 2014, http://www.who.int/ith/mode_of_travel/cab/en/, accessed 19 March 2014.
[14] Oxygen: with or without oz, Mounteverest.net, http://www.mounteverest.net/expguide/wwoz.htm, 19 March 2014.
[15] Flight level, Wikipedia, 25 February 2014, http://en.wikipedia.org/wiki/Flight_level, accessed 19 March 2014.
[16] Noctilucent cloud, Wikipedia, 27 February 2014,

Measuring the World, by John Austin

http://en.wikipedia.org/wiki/Noctilucent_clouds, accessed 19 March 2014.
[17] Aurora, Wikipedia, 17 March 2014, http://en.wikipedia.org/wiki/Aurora, 19 March 2014.
[18] Karman line, Wikipedia, 12 March 2014, http://en.wikipedia.org/wiki/K%C3%A1rm%C3%A1n_line, accessed 19 March 2014.
[19] How much suction in a vacuum cleaner?, M. Green, Blogspot, 27 January, 2013, http://marty-green.blogspot.co.uk/2013/01/how-much-suction-in-vacuum-cleaner.html, accessed 19 March 2014.
[20] Ishimaru, H., 1989: Ultimate Pressure of the Order of 10^{-13} torr in an Aluminum Alloy Vacuum Chamber, *Journal of Vacuum Science and Technology,* **7** (3–II), 2439–2442. doi:10.1116/1.575916, accessed 19 March 2014.
[21] Vacuum, Wikipedia, 11 March 2014, http://en.wikipedia.org/wiki/Vacuum, accessed 19 March 2014.
[22] Barometer, Wikipedia, 13 March 2014, http://en.wikipedia.org/wiki/Barometer, accessed 19 March 2014.
[23] Vernier scale, Wikipedia, 3 March 2014, http://en.wikipedia.org/wiki/Vernier_scale, accessed 19 March 2014.
[24] Atmospheric pressure, UK Meteorological Office Observer's handbook (Met.O.1028) 4th Edition, Chapter 7, reprinted 2000, http://www.metoffice.gov.uk/archive/observers-handbook-4th-edition-reprint, accessed 19 March 2014.
[25] Trillionthtonne.org, explaining the need to limit cumulative emissions of carbon dioxide, http://trillionthtonne.org/, accessed 19 March 2014.
[26] Trends in atmospheric carbon dioxide, NOAA Earth Systems Research Lab., Global Monitoring Division, March 2014, http://www.esrl.noaa.gov/gmd/ccgg/trends/, accessed 19 March 2014.
[27] Fifth assessment report, Intergovernmental Panel on climate change, http://www.ipcc.ch/index.htm#.Uym0DpWPPIU, accessed 19 March 2014.
[28] Stern Review, Wikipedia, 16 March 2014, http://en.wikipedia.org/wiki/Stern_Review, accessed 19 March 2014.
[29] Coriolis effect, Wikipedia, 18 March 2014, http://en.wikipedia.org/wiki/Coriolis_effect, accessed 19 March 2014.
[30] Buys Ballot's law, Wikipedia, 17 October 2013, http://en.wikipedia.org/wiki/Buys_Ballot's_law, accessed 19 March 2014.

[31] Sea breeze, Wikipedia, 4 March 2014, http://en.wikipedia.org/wiki/Sea_breeze, accessed 19 March 2014. Also, image by Jesus Gomez Fernandez, 19 June 2005, http://commons.wikimedia.org/wiki/File:Diagrama_de_formacion_de_la_brisa-breeze.png?uselang=en-gb, accessed 19 March 2014.
[32] Deep diving, Wikipedia, 20 March 2014, http://en.wikipedia.org/wiki/Deep_diving, accessed 20 March 2014.
[33] Ascending from a dive, B.R. Morris, 2011, http://www.diverssupport.com/safetystops.htm, accessed 20 March 2014.
[34] Oxygen narcosis, Scubaboard.com, 30 December 2002, http://www.scubaboard.com/forums/basic-scuba-discussions/19820-oxygen-narcosis.html, accessed 20 March 2014.
[35] Nitrogen narcosis, Wikipedia, 27 January 2014, http://en.wikipedia.org/wiki/Nitrogen_narcosis, accessed 20 March 2014.
[36] Trimix (breathing gas, Wikipedia, 31 January 2014, http://en.wikipedia.org/wiki/Trimix_(breathing_gas), accessed 20 March 2014.
[37] Sperm whale, Wikipedia, 18 March 2014, http://en.wikipedia.org/wiki/Sperm_whale, accessed 20 March 2014.
[38] Deep sea creature, Wikipedia, 15 February 2014, http://en.wikipedia.org/wiki/Deep_sea_creature, accessed 20 March 2014.
[39] Structure of the Earth, Wikipedia, 17 March 2014, http://en.wikipedia.org/wiki/Structure_of_the_Earth, 20 March 2014.
[40] Sound, Wikipedia, 19 March 2014, http://en.wikipedia.org/wiki/Sound, accessed 20 March 2014.
[41] List of scientists whose names are used as SI units, Wikipedia, 16 November 2013, http://en.wikipedia.org/wiki/List_of_scientists_whose_names_are_used_as_SI_units, accessed 20 March 2014.
[42] 10 limits to human perception …. and how they shape your world, R.T.Gonzalez, IO9, 17 July 2012, http://io9.com/5926643/10-fundamental-limits-to-human-perception----and-how-they-shape-your-world, accessed 20 March 2014.
[43] Psychoacoustics, Wikipedia, 26 February 2014, http://en.wikipedia.org/wiki/Psychoacoustics, 20 March 2014.

[44] Hearing range, Wikipedia, 11 March 2014, http://en.wikipedia.org/wiki/Hearing_range, accessed 20 March 2014.
[45] Animal echolocation, Wikipedia, 17 March 2014, http://en.wikipedia.org/wiki/Animal_echolocation, accessed 20 March 2014.
[46] Sound, Infoplease, Dorling Kindersley online encyclopedia, 2007, http://www.infoplease.com/dk/encyclopedia/sound.html, accessed 20 March 2014.
[47] Equal-loudness contour, Wikipedia, 4 February 2014, http://en.wikipedia.org/wiki/Equal-loudness_contour, accessed 21 March 2014.
[48] Atmospheric waves, M. Miller, Course notes, European Centre for Medium Range Weather Forecasts, April 1987, http://www.ecmwf.int/newsevents/training/rcourse_notes/NUMERICAL_METHODS/ATMOSPHERIC_WAVES/Atmospheric_waves5.html, accessed 21 March 2014.
[49] Skamarock, W.C. & J.B. Klemp, 1993: Adaptive grid refinement for two-dimensional and three-dimensional nonhydrostatic flow, Monthly weather Review, 121, 788-804, http://www.mmm.ucar.edu/people/skamarock/Papers/cv_07.pdf, accessed 21 March 2014.
[50] Speed of sound, Wikipedia, 19 March 2014, http://en.wikipedia.org/wiki/Speed_of_sound, accessed 21 March 2014.
[51] Mach number, Wikipedia, 20 March 2014, http://en.wikipedia.org/wiki/Mach_number, accessed 21 March 2014.
[52] Meteor hits Russia 2013 – best footage collection, Youtube, 18 February 2013, http://www.youtube.com/watch?v=dpmXyJrs7iU, accessed 21 March 2014.
[53] Doppler effect, Wikipedia, 20 March 2014, http://en.wikipedia.org/wiki/Doppler_effect, accessed 21 March 2014.
[54] Doppler radar, Wikipedia, 19 March 2014, http://en.wikipedia.org/wiki/Doppler_radar, accessed 21 March 2014.
[55] Speeds of sounds of the elements, Wikipedia, 17 February 2014, http://en.wikipedia.org/wiki/Speeds_of_sound_of_the_elements_(data_page), accessed 21 March 2014.
[56] The speed of sound and elastic constants, Introductory material, http://iowadoppler.com/documents/speed-manual.pdf, accessed 21

March 2014.
[57] Speed of sound in some common solids, Engineering Toolbox, http://www.engineeringtoolbox.com/sound-speed-solids-d_713.html, accessed 21 March 2014.
[58] Earthquake, Wikipedia, 21 March 2014, http://en.wikipedia.org/wiki/Earthquake, accessed 21 March 2014.
[59] April 2011 Fukushima earthquake, Wikipedia, 18 March 2014, http://en.wikipedia.org/wiki/April_2011_Fukushima_earthquake, accessed 21 March 2014.
[60] See what 1000 tiny fracking caused earthquakes can do to a home, the AtlanticCities.com, 29 August 2013, http://www.theatlanticcities.com/neighborhoods/2013/08/see-what-1000-fracking-caused-earthquakes-can-do-home/6712/, accessed 21 March 2014.
[61] Richter magnitude scale, Wikipedia, 8 March 2014. http://en.wikipedia.org/wiki/Richter_magnitude_scale, accessed 21 March 2014.
[62] The modified Mercalli intensity scale, US Geological Survey, http://earthquake.usgs.gov/learn/topics/mercalli.php, accessed 21 March 2014.
[63] Measuring tyre footprint with the TireScan, Youtube, 20 December 2010, http://www.youtube.com/watch?v=Jkp9UiIA-0A, accessed 21 March 2014.
[64] How mufflers work, K. Nice, Howstuffworks.com, 19 February 2001, http://auto.howstuffworks.com/muffler.htm, accessed 21 March 2014.
[65] Catalytic converter, Wikipedia, 18 March 2014, http://en.wikipedia.org/wiki/Catalytic_converter, accessed 21 March 2014.

11. Work, Energy and Power
Work = the movement of an object against a resisting force
Energy = the capacity for doing work
Power = the rate of doing work

11.1 Introduction

Work, energy and power are important physical concepts, and so definitions for all three are given above. To oppose a *force* requires *work* and to sustain work requires *energy*. For example, to lift a weight requires doing work against the resisting force of gravity. You can think of energy as a reservoir which allows work to be done, and both work and energy are measured in the same units. *Power*, the rate of doing work, is the work done per unit time and is frequently important to every day life. For example, we measure the power of our car engines, and the power needed to heat our homes.

Heat and mechanical energy were originally thought to be separate quantities, but we now know them to be different forms of energy in general, as discussed in § 11.2 and § 11.3. Consequently, the metric system uses the same units for both. In fps, the units are often confusing, and there is a separate set of units for heat energy (therms). Moreover, the situation is even worse since heat energy units don't have a single SI energy equivalent. Rather, a number of exact equivalents seem to be in circulation so you may have to guess as to which one you're using.

Heat energy is very closely related to dietary energy as

explained in § 11.4 and so food energy is often still quoted in calories or *kcal*, dating to the time when heat and mechanical energy were treated separately. This unit is rarely used now outside culinary science. Consumption of food is essentially a chemical process and the energy changes produces heat which is measured and recorded in calories. Countries that use fps also tend to choose the calorie for dietary energy, producing a mishmash of unit systems. Power is introduced in § 11.5, and exercise, the flip side of dietary energy, is discussed in § 11.6.

Very large energy expenditures (§ 11.7) are often compared with the explosive capacity of trinitrotoluene ($CH_3C_6H_2(NO_2)_3$, TNT), which is something of an anachronism, preventing easy comparison with more familiar units of energy. Proper energy units would certainly be more informative to me. The energy and power of natural processes such as thunderstorms and volcanoes are often described in bomb terms, but the comparison is often misleading as we discover in § 11.8. The Chapter ends with a discussion of machines and the fuels to drive them.

Electrical energy and power needs a completely new set of supporting units for current, voltage and so on. These are discussed in Chapter 12. Amazingly, the same electrical units are used worldwide, an important development suggesting that agreement on all units may yet be possible. At the moment, fps countries (mainly the USA) are very confused in relating one type of energy to another, be it mechanical, dietary, chemical or electrical.

Incidentally, the word 'energy' is frequently abused by faith healers and related fringe practitioners. Whenever you see the word, ask yourself what the associated force is, and whether it is recognised by science (§ 9.1).

Measuring the World, by John Austin

11.2 Mechanical versus heat energy

Mechanical energy is the energy needed to oppose a mechanical force. Heat energy is the energy needed to raise the temperature of an object. The amount of mechanical energy is the force multiplied by the distance moved by the force.

In the metric system, work and energy are measured in joules (J) = Nm (SI system) or ergs = dyn cm. The latter is a rather small unit, obtaining by multiplying a small force (1 dyne) by a small displacement (1 cm). Since 1 dyne = 10^{-5} N, 1 erg = 10^{-5}N x 10^{-2} m = 10^{-7} J. The unit of heat energy is the calorie, initially defined as the energy required to raise the temperature of 1 g of water by 1 °C. More modern physics defines the specific heat capacity (SHC) as the energy required to raise the temperature of a given material by 1 °C per unit mass[1]. In SI units, the SHC of water varies slightly with temperature from about 4182 at 20 °C to 4204 $Jkg^{-1}K^{-1}$ at 4 °C[2], and there is possibly a still wider range at other temperatures. Because of this variation, the calorie came to be described as the '15 °C calorie' (the SHC between 14.5 and 15.5 °C), as well as other variants. This came to be quite messy, and so for precise work, it was convenient to define the calorie at exactly 4.184 J but there is still some confusion with 4.186 J sometimes used[3].

The fps system has also has some antiquated units. Work is strictly measured in ft pdl, which can be confused with ft lb. If the latter is used for work or energy, it usually means that g, gravity, needs to be included as an additional factor. Thus 10 ft lb = 322 ft pdl (approx.). In using the fps system, the search for hidden factors of g causes me a serious headache! There is no danger with the metric system as I have never seen the equivalent, kg m, to define work. Unfortunately, some people who are already confused by the fps system try to spread that confusion to the metric system. So, for example, tyre pressure gauges are marked with kg cm^{-2}, the direct analogy of psi as noted in Chapter 10. The concept of using water to define energy as we saw for the calorie also applies for the fps system. The BTU, the British (oh, the shame!) thermal unit is defined as the energy needed to raise one pound of water by 1 °F. Since this is a small unit, the therm (10^5 BTU) is often used instead. The imagination (or is it just plain disorganisation?) of the fps system reveals itself again: why 10^5 and not say 10^4, or 10^6, or 10^7?

Therms are typically used on US gas utility bills, whereas electricity consumption (Chapter 12) is usually given in kW h. This is a product of power (kW) and duration (hours). So sometimes in the USA you can get charges in different units on the same bill, if the power company supplies both gas and electricity. Many years ago, the UK also charged for energy consumption by gas in therms and electricity in kW h, making them difficult to compare easily. For example, you might have the choice of heating your home by the more convenient electricity, or with gas. In fact gas has a significant pricing advantage compared with electricity, so it makes sense from the perspective of the gas company to quote prices in kW h. This is what they now do in the UK[4].

11.3 Conversion between units

To convert 1 BTU to J, we use the conversion rates of 1 °F = 5/9 K, 1 lb = 0.45359237 kg and the specific heat capacity of water. However, just like the calorie, nobody seems to be able to agree on which specific heat capacity to use, so precise values vary by a few parts in 1,000[5], depending on source. This is unsatisfactory. The 'thermochemical' BTU uses the 4.184 J calorie, which I'll use here:

1 BTU = 0.55555 x 0.45359237 x 4184 = 1054.350 J

The BTU varies in size between 1054.35 and 1059.67

J[5], but is mostly below 1056 J. An approximate conversion from calories to joules is to multiply by 4 and add 5% (error 0.4%). To convert from J to cal divide by 4 and subtract 5% (error 0.4%). To convert from BTU to J, multiply by 1000 and add 5% (error 0.4%) and to convert from J to BTU, divide by 1000 and subtract 5%. Of course a simpler conversion from BTU to J and the converse is to multiply or divide by 1000, giving an error of about 5.4%, which saves a lot of arithmetic and may be quite adequate for most purposes.

11.4 Dietary Energy

To obtain its calorie content, Food may be treated like any other material, and burnt in a calorimeter[6], to measure the heat produced. Corrections are then made to the values obtained to take into account the energy required for digestion and absorption[7,8] to compute the dietary value of a foodstuff. Evidence suggests that there is little waste energy excreted in the urine or faeces[7,8], but determining exact energy intake by individuals has proved to be difficult even under carefully controlled conditions. However, there are some exceptions in that for some materials such as cellulose, we don't possess the appropriate bacteria or enzymes, whereas other animals (e.g. cows) may do.

Traditionally, dietary energy is recorded in calories even in the USA where otherwise the metric system is essentially unknown. The abbreviation Cal (capital C) is often used as shorthand for kcal. However, I think the kcal abbreviation is becoming common, so I will preferentially use it. So, 1 kcal = 4.184 kJ. Normal daily requirement for a sedentary adult is about 10,500 kJ (2500 kcal) for a man and 8400 kJ for a woman (2000 kcal)[9]. The *kJ* is becoming more familiar in the labelling of food, even though in Europe at least it appears alongside the kilocalorie amount. While a dual system is in existence, there is no incentive for the consumer to switch to the proper energy unit. Sometimes the best strategy to change units is to go 'cold turkey' (an interesting food analogy!) and simply to stop writing the old units on packaging, forcing people to flounder (sic!) around a little before they understand the new units. That strategy applies to the metric system as a whole, of course, not just to dietary units.

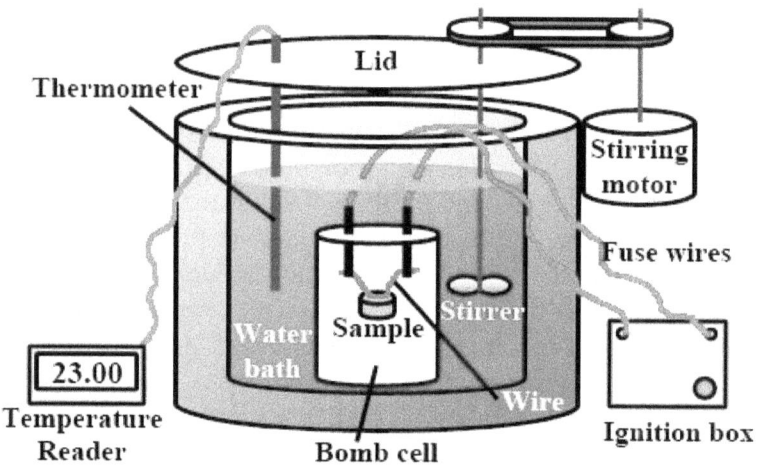

Diagram of a bomb calorimeter, used to determine the dietary energy of food[10]. From Wikimedia, author: Lisdavid89, 4 November 2012.

11.5 Power Units

The SI unit for power is the watt[11], W, named after the Scottish engineer James Watt. This is the rate of doing work: 1 $W = 1\,J\,s^{-1}$. The metric system treats all forms of power in the same units and almost everyone is familiar with electrical power, which is discussed in Chapter 12. Several fps power units are used by specialist groups. These include, strictly, ft pdl s^{-1}. The unit ft lb s^{-1} is similar but a factor g = 32.2 ft s^{-2} larger. Neither of these units is in common use. The most used fps power unit is the horse power (hp), presumably based on some idealised horse, with 1 hp = 550 ft lb s^{-1} (exactly) = 550 x 0.3048 x 0.45359237 x 9.80665 = 745.69987 W, where 0.3048 converts ft to m, 0.453... converts lb to kg and 9.8... is the ubiquitous g factor because of the use of lb instead of poundals. So, 1 hp = 745.70 W and 1 kW = 1/0.74570 = 1.34102209 hp. This is known as the mechanical horse power. There are also electrical (746 W exactly) and hydraulic horse powers, which both differ slightly from the mechanical horse power: another reason not to use it.

Typical power ratings[12] vary from the minute (10^{-21} W – the power of Galileo space probe radio signal) to the astronomical (10^{49} W – the total luminosity of all the stars in the observable universe). Power levels that we are more familiar with include the output from car engines, which is typically in the range 40 – 200 kW. This is often quoted as a horsepower, about 50 – 270 hp. It is difficult to visualise a cart being pulled by 50 horses, but a sleigh pulled by 12 dogs makes a great sporting activity such as the Iditarod in Alaska[13]. The dogs love to run and probably generate 300 – 400 W (~½ hp) of useful power each. Power is often expressed in brake horsepower, which is the available power after correcting for friction and other engine losses, e.g., in an engine. The SI system uses less jargon, and refers to input energy and output energy or power from an engine.

11.6 Energy Expenditure of Exercise

Energy consumption during exercise is an important issue for many people, as it can provide the basis for personal weight control. In humans, exercise is only about 18 - 26% efficient[14], meaning that for the useful work done, perhaps by cycling, or rowing, about 74 – 82 % of the energy needed goes to heating the muscles and driving the cardiovascular system. This relative inefficiency needs to be taken into account when computing the energy consumed for weight control or dietary requirements from first principles. Energy usage by exercise machines are usually calibrated to show this almost factor of four between work done and input energy to give the equivalent dietary requirement. The following table shows an estimate of the energy (dietary) requirement[15,16] for a range of activities, converted from funny units to kJ min^{-1} kg^{-1}. The walking and running figures are all taken from [16] which shows an extensive set of data at different speeds, and generally quoted to higher precision.

Energy need (kJ per minute) for selected activities per kg of body mass.

Activity	Energy need	Activity	Energy need
Aerobics (high impact)	0.65	Swimming (free)	0.55
Aerobics (low impact)	0.4	Swimming (breast)	0.65
Cycling 16 km/h	0.55	Swimming (back)	0.7
Cycling 28 km/h	0.8	Tennis match	0.7
Football	0.7	Walking 5.6 km/h	0.27[16]
Golf (carrying clubs)	0.4	Walking 8 km/h	0.56[16]
Gymnastics	0.3	Running 8 km/h	0.64[16]
Strength training: circuit	0.55	Running 16 km/h	1.20[16]

The comparison of walking and running expenditures[16] is instructive, so will be explored further,

Measuring the World, by John Austin

particularly as the values are available to higher precision than the other activities. For a typical person, energy per unit distance travelled minimises for walking at about 5 km h^{-1}, and is higher for running. This makes sense since walking efficiency perhaps peaks at a relaxed pace, and then the energy expenditure increases as the walking speed increases. Also, the running efficiency is consistently lower than walking, which makes sense as running is more difficult than walking! It is somewhat intriguing that energy expenditure of running decreases on a per distance basis as the speed increases. Nonetheless, energy use at 5.6 km h^{-1} walking (3.5 mph) for a 70 kg person covering 1 km equates to 0.27 x 60/5.6 x 70 = 202 kJ, or 202x1.6/4.184 = 77 kcal per mile. The 70 kg runner at 16 km h^{-1} is consuming 1.20 x 60/16 x 70 = 315 kJ, or 315x1.6/4.184 = 120 kcal per mile.

Running at 7 min miles, which is about my limit these days, 1 mile would take 420 s, and according to[16] the energy consumption is 80 kcal per 100 pounds per mile, or assuming my weight of 65 kg (143 lb), 80x1.43 = 114.4 kcal per mile. I am using a mixture of units here, not by choice, but because[16] uses this mixture and it is easier to convert to metric at the end of the calculation. Hence, the energy consumed would be 479 kJ in 420 s, or 1.14 kJ s^{-1}, or 1140 W. The useful work would be about 20% of this, or 230 W, or 0.3 hp. It is reassuring that our putative (very mediocre) athlete is somewhat less athletic than our primary standard horse! Cyclists at the top level often meter their power output, and their typical levels are 450 W (6.4 W kg^{-1}) for a man and 300 W for a woman (5.7 W kg^{-1}) sustainable for an hour[17] to generate a time trial speed of 40 km h^{-1} on a flat road (for a male). For dietary purposes, energy consumed would be about 4 times this. In the table, the energy consumed at 28 km h^{-1} is 800 J min^{-1}kg^{-1}, or 800/60 W kg^{-1} = 13.3 W kg^{-1} or about 3.33 in useful power output. Cycling power works at the square of the speed, so at 40 km h^{-1} (25 mph) our 'recreational cyclist', assumed to be male for the purpose of illustration, would instead be generating about 3.33x40^2/28^2 of useful power = 6.8 W kg^{-1}. This is quite close to the 6.4 quoted, so everything is consistent. In all these situations, it is important to recognise the difference between useful power generated and the dietary requirement of a particular exercise, which is likely to be four times higher.

Human powered flight has been made possible in recent years by improved understanding of aerodynamics and the development of strong, lightweight materials. To fly a craft, typically a bicycle pedal is used to power a rotor which provides forward speed. Airflow over the wings provides lift. With such a machine, built as early as 1977, flight could be achieved with a total power output of only 260 W[18], which is the standard of a 'very good' competitive cyclist sustainable for an hour[17]. That is for every kg airborne (which would need to include human and craft), only about 3 W power is needed. This compares with bird flapping flight that are at least 40 W kg^{-1}[19]. Gliding flight has a low power requirement (4 W kg^{-1})[20], similar to human powered machines. With the much lower power requirements of human-powered machines, compared with bird flapping flight we lose the manoeuvrability skills that are essential to birds and insects.

11.7 Very Large Energy Units : Bombs

The metric system does not distinguish between high and low energy units: joules, with the appropriate pre-multiplier is used for both. A commonly used unit in fps for bomb yields is the explosive energy of 1 ton of TNT. This is fixed by convention at 4.184 GJ, i.e. 1 Gigacalorie[21]. Actually a tonne of TNT yields 4.686 GJ. The atom bomb (A-bomb) on Hiroshima for example gave a yield of about 15 kton (63 TJ). Modern fusion devices (H-bombs) are even more powerful. In a hydrogen bomb, a major release of energy comes from an initial implode triggered by a fission bomb (A-bomb), which raises the temperature of the H-bomb core to over 10 million K [21, 22]. Older bombs release the energy of 4000 Hiroshima-sized bombs (about 250 PJ). Modern nuclear devices have a smaller yield (for strategic purposes) of under 5 PJ, with most now in the range 8 – 40x Hiroshima = 500 – 2500 TJ[24]. Detonation of a nuclear weapon passes through several stages[25]. The blast stage is an overpressure of the atmosphere which destroys most buildings within a mile radius and moderate to severe damage further away. The atmospheric pressure gradient generates strong winds, approaching the speed of sound several times that of a hurricane. The majority of people a short distance from the blast are killed by a lethal dose of radiation, with further deaths and injuries beyond. The initial stage of thermal radiation: UV, visible and Infra red causes flash burns, and at Hiroshima and Nagasaki caused 20 – 30% of the deaths. Smoke and dust from fires generate atmospheric pollution which blocks out the sun and lowers the temperature with the risk of nuclear winter. The expected effects vary in detail depending on bomb yield and atmospheric conditions.

Note that in the above account, we have used the metric pre-multipliers G (10^9), T (10^{12}) and P (10^{15}) for the humble-sized joule. If it weren't for the ton of TNT, which is a type of metric unit anyway (based on heat energy), then the fps would have to use some sort of silly unit like horse power year or therms or quads. An approximate (5% error) conversion is 1 PJ = 1×10^{12} therms = 0.001 quads (for the definition of quad, see § 12.9). When fps can't cope, it uses a vague metric unit!

11.8 Natural Phenomena

Energies and powers of natural systems such as earthquakes, hurricanes and weather systems typically dwarf human produced power. For example, there is strong geological evidence that the mass extinction which occurred 65 million years ago, which included the loss of the dinosaurs, was due to an asteroid impact of about 10^8 megatons of TNT[26], or 4×10^{23} J, or 6×10^9 x Hiroshima. This date is known as the K-T boundary, short for Cretaceous-Tertiary, the two geological eras, after the German word Kreidezeit. The energy released is often expressed as 10^8 megatons TNT, making an emotive comparison with explosives, but for other processes this is not always helpful.

For example, the energy of a tropical storm is of order 36 GJ (8.6 Mt TNT)[21], comparable with a low-yield weapon. However, the latter acts on a very small spatial scale (order km) whereas the former acts on a very large spatial scale (order 1000 km), so naturally the effects are very different. The energy released by a hurricane in a day is even larger at 5×10^{19} J (50,000 PJ or 12,000 megatons TNT)[27]. Volcanic eruptions can also be compared with the energy of explosions[28]. For example, the large eruption of Krakatoa in 1883, was noted as the equivalent of 200 megatons TNT (800 PJ). With this comparison, the tendency is to think of a volcano as a large bomb, and again this is not always appropriate because of the additional material released to the atmosphere, including CO_2 and SO_2. However, in comparison with a weapon, the energy release of a volcano is more like a slow fuse, with many of the atmospheric effects appearing months later. Krakatoa was a somewhat rare volcanic eruption, and more typical ones have smaller energy releases, by orders of magnitude.

Another very large natural phenomenon is an earthquake. At the Earth's surface, these have energy releases comparable to volcanoes, e.g. 1×10^{17} J for the 2004 Indian Ocean earthquake[28] at the Earth's surface, yet as much as 4×10^{22} J for the total energy. Clearly, like volcanoes, every earthquake will be different. However you look at it, these are very large energy releases, and expressing the figures in megatons of TNT hardly seems to convey any useful information. Rather, it complicates the situation in my view, as both metric and closet metric units are often employed together with all sorts of comparisons with human engineering feats, be it big bombs or annual electric or national power station output. I find after a while this just gives me a headache trying to keep track of all the energies, and I would personally much prefer to see just the energy number, in proper units of course!

11.9 The performance of Aircraft and Rockets

One of the growing sectors of energy consumption is aviation. In 1992, aviation produced about 2% of the world's greenhouse gas emissions[29], small compared with other sources. However, aviation also deposits water vapour directly in the upper troposphere and lower stratosphere (about 10 km altitude) where commercial jet aircraft cruise. This is visible through condensation trails. The extent to which they affect climate is likely to be very small, but it is growing [29,30]. Overall, flying can be an energetically efficient form of travel. For example, an 8000 km (10 hour) flight of a jet airliner such as a Boeing 747 requires about 36,000 US Gallons of fuel[31] (136,000 L) to transport, say, 500 people including baggage. This works out at about 8000x500/136,000 = 29 km per person per litre of fuel. In my UK car, I typically get about 14 km L^{-1}. So flying is twice as efficient as driving with single occupancy.

Jet propulsion works on the principle of burning fuel with compressed air from the atmosphere surrounding the aircraft[32]. At high altitude, atmospheric pressure is insufficient to support sufficiently rapid combustion. This puts a natural limit to the flying altitude at about 20 km. By expelling gases at high temperature and velocity, forward thrust is supplied by the engines. In practice, commercial aircraft fly at considerably lower altitudes – about 9 km. The combustion temperature could reach almost 1400 °C[33], but engines are cooled with outside air. Commercial jet engines have been built with thrusts of up to 558 kN for the GE90-115B, which is specifically designed for the Boeing 777[34].

A rocket carries its own oxidant, as well as the fuel, collectively known as propellants, and as in the case of jet engines, the expelled exhaust gases provide forward acceleration[35]. Rockets for space use are usually staged: material is jettisoned after the fuel within the stage is exhausted, to provide a smaller mass and hence greater acceleration to the remaining stages and payload. Because it carries its own oxidant, a rocket functions as well in space as within the atmosphere. Rockets are essentially explosives with a controlled burn, usually to impart momentum and deliver a payload. The Saturn V rocket[36], was perhaps one of the icons of rocket history, having launched the Apollo spacecrafts to the moon. Over 2,000 t of propellants were consumed in stage 1, providing 34 MN of thrust during a flight time of 150 s. The rocket design for the space shuttle[37] was very different due to the requirement of reusing as much as possible of the equipment. The first stage consisted of two solid fuel rockets providing 25 MN thrust in total for the first two minutes of flight. Another 4 MN was provided by the shuttle main engines using propellants of liquid H_2 and O_2. As well as high temperature in the combustion chamber, high pressure (exceeding 20 MPa, 200 atmospheres) was also experienced.

Fireworks are similar in principle to rockets, without the payload[38]. The ignition materials consist of gunpowder and other ingredients. Coarse gunpowder, tightly packed, propels rockets. A finer, more closely packed, gunpowder explodes the rocket once it is airborne. Colours are obtained by including various salts in the mix: sodium (yellow), strontium (red), copper (blue) and barium (green). Charcoal is added to create a sparkling, flaming tail.

11.10 Energy Content of Fuels

The choice of a particular fuel for a particular task depends on a number of factors including the energy available, the suitability of using a particular fuel, and its safety. For example, to make an obvious point, jet aircraft need liquid fuel so that the gaseous exhaust can provide forward thrust. Electric power stations (Chapter 12) are flexible and can be adapted to run on solid fuel (e.g. Coal) or liquid fuel (e.g. Oil or liquefied gas).

The most important fuels are coal, oil and natural gas. Crude oil is often found in conjunction with natural gas. Crude oil is refined to extract different density oils and fuels for heating and transport. Natural gas consists mostly of CH_4 with C_2H_6 (ethane), C_3H_8 (propane) and butane (C_4H_{10}) in steadily decreasing amounts and with some N_2 and CO_2 in its impure state. The alkanes above, and also pentane and hexane are obtained from liquefying natural gas. The energy released from natural gas is about 54 MJ kg^{-1}[40], or a value in the range 37-40 MJ m^{-3} (depending on the CH_4 content. This energy is sometimes expressed as 1047-1130 J ft^{-3} or 0.99-1.07 BTU ft^{-3}. Liquid H_2 is a rocket fuel and is too explosive to be considered safe in general use as a liquid, but on a per mass basis has the highest heat of combustion at 142 MJ kg^{-1}. However, the density is about 8 times lower than CH_4, so on a volume basis, the energy from burning is much lower than natural gas. Nonetheless, hydrogen powered cars have been suggested as an alternative to gasoline[41]. In principle hydrogen can be made safe by putting it into a matrix with other material, although the technology is not fully developed.

Ethyl alcohol (ethanol) is sometimes used as a fuel in car engines, but it is corrosive and engines need to be specially adapted. The advantage is that ethanol can be fermented from plants with high sugar content, as is done e.g. In Brazil (sugar cane) and USA (corn). Ethanol mixes with gasoline, so mixtures are often sold in petrol stations. E15 (15% ethanol) can be used by modern car engines without adaptation. Although ethanol gives only 63% of the energy of gasoline, by mass[40], it can be an important strategy for reducing oil consumption and net greenhouse gas emissions. Although there is still debate about the environmental impact, the CO_2 absorbed by plant growth is approximately matched by that produced by plant decay and the burning of the alcohol produced[42].

11.11 Simple Machines

The purpose of a machine is to convert one form of energy into another, or to reduce the force required to do a particular task, or to obtain a speed advantage[43]. A common measure is the energy efficiency (EE), given by work done divided by energy consumed. For a motor EE is about 75%, and as high as 98% for an electrical transformer in AC (Chapter 12). Biological species have much lower efficiencies, about 24% for humans (§ 11.6), but they are machines nonetheless, transforming chemical energy to other forms. Other measures of machines are their mechanical advantage, MA = force obtained / force applied, and velocity advantage, VA = velocity obtained / velocity applied. By the definition of energy production and conservation, $EE = MA \times VA$. Hence $MA > 1$ implies $VA < 1$ and vice versa. A bicycle is a machine designed to give a velocity advantage ($VA > 1$). A lever (depending on its design) is a machine designed to give a mechanical advantage ($MA > 1$). Machines can be reduced to a set of simple machines[43].

Although we don't think of them in this way very often, simple machines in everyday use include the inclined plane,

Measuring the World, by John Austin

levers, pulleys and of course the wheel. With an inclined plane, an object is moved gradually up a slope rather than being lifted directly.

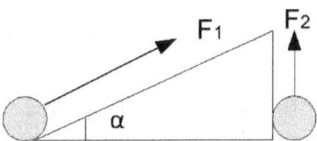

In the above diagram, the force to lift the object a height h is F_1 and the distance along the plane is $h/\sin\alpha$, where α is the angle of the plane. So the work done in lifting the object is $F_1 h/\sin\alpha = F_2 h$. Hence, $F_1 = F_2 \sin\alpha$. The mechanical advantage is $F_2/F_1 = 1/\sin\alpha$ and the velocity advantage is $\sin\alpha$. For example, for a slope of 1 in 10, $\sin\alpha = 0.1$ and $F_1 = 0.1 F_2$, a considerable force reduction or mechanical advantage. In practice, the plane could be highly efficient but instead of the force F_1, there would need to be an additional force F_x to counteract friction, and $F_1 = F_x + F_2 \sin\alpha$. If the object is metallic, and the plane is also made of metal, the coefficient of sliding friction is about 0.02, making the plane 98% efficient.

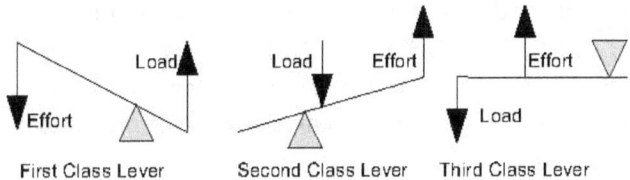

First Class Lever   Second Class Lever   Third Class Lever

Measuring the World, by John Austin

The lever is another important simple machine and may be divided into three particular types, known as *classes*, as in the figure. An example of a first class lever is that used in a warehouse to move a heavy load with a smaller effort. The wheelbarrow is an example of the second class of lever, in which case the load is close to the fulcrum, about which the lever pivots. An example of the third class of levers is the human arm, which provides a mechanical disadvantage but a velocity advantage. The forces in any particular case can be calculated by balancing the product of effort and distance from the fulcrum with the product of distance and load from the fulcrum. The quantity obtained is the *torque* about the fulcrum and is measure in N m or (preferably not) ft lb. The invention of the lever has been attributed to Archimedes who in his (over) enthusiasm said that "Give me a long enough lever and a fulcrum on which to place it, and I shall move the world"[44]. But for fun, let's see if this holds up to scrutiny! We need the first class lever for which the load is the mass of the Earth times an acceleration. The effort would be what a human could expend, but let's be generous and say 200 g N ≈ 2000 N. If x is the distance from the fulcrum, then we have $2000x = 6.0 \times 10^{24} \times 10^{-3} \times 10^{-3}$. In this calculation, the mass of the Earth is taken from § 9.7. I have also assumed a distance of 1 mm for the Earth from the fulcrum (impossibly small!) and 1 mm s^{-2} (quite minuscule) for the acceleration of the Earth from its mean orbital path. Solving the equation, we find $x = 3 \times 10^{15}$ m approximately, or the distance travelled by light at 3×10^8 m s^{-1}, of 10^7 s or 100 days. So, our lever would need to be at least ¼ of a light year long! This is of course an impossibly large lever but the calculation has provided a bit of entertainment!

Other simple machines include the wheel and axle[45] and pulleys[46]. Similar to levers, they can both be used to provide a mechanical or velocity advantage when accompanied with gears.

Measuring the World, by John Austin

The reader is encouraged to explore the references[45,46] to understand these simple machines in more depth.

References

[1] Heat capacity, Wikipedia, 16 March 2014, http://en.wikipedia.org/wiki/Heat_capacity, accessed 22 March 2014.
[2] Calorie, Wikipedia, 20 March 2014, http://en.wikipedia.org/wiki/Calorie, accessed 22 March 2014.
[3] Hargrave, J.L., 2007: Does the history of food energy units suggest a solution to "Calorie confusion"?, Nutrition Journal, 6:44, doi:10.1186/1475-2891-6-44, http://www.nutritionj.com/content/6/1/44, accessed 22 March 2014.
[4] Helping you to understand your British Gas bill, Which, 12 January 2013, http://www.which.co.uk/documents/pdf/british-gas-bill-323865.pdf, accessed 22 March 2014.
[5] British thermal unit, Wikipedia, 18 March 2014, http://en.wikipedia.org/wiki/British_thermal_unit, accessed 22 March 2014.
[6] Calorimeter, Wikipedia, 20 March 2014, http://en.wikipedia.org/wiki/Calorimeter, accessed 22 March 2014.
[7] Food energy, Wikipedia, 1 March 2014, http://en.wikipedia.org/wiki/Food_energy, accessed 22 March 2014.
[8] Atwater system, Wikipedia, 29 January 2014, http://en.wikipedia.org/wiki/Atwater_system, accessed 22 March 2014.
[9] What should my daily intake of calories be?, NHS choices, http://www.nhs.uk/chq/pages/1126.aspx?CategoryID=51&SubCategoryID=165, accessed 22 March 2014.
[10] Bomb calorimeter diagram, Wikimedia, Lisavid89, 4 November 2012, http://commons.wikimedia.org/wiki/File:Bomb_Calorimeter_Diagram.png?uselang=en-gb, accessed 22 March 2014.
[11] Watt, Wikipedia, 14 March 2014, http://en.wikipedia.org/wiki/Watt, 22 March 2014.
[12] Orders of magnitude (power), Wikipedia, 25 February 2014, http://en.wikipedia.org/wiki/Orders_of_magnitude_(power), accessed 22 March 2014.

[13] Iditarod, The last great race, http://iditarod.com/, accessed 22 March 2014.
[14] Muscle, Wikipedia, 16 March 2014, http://en.wikipedia.org/wiki/Muscle, accessed 22 March 2014.
[15] How to calculate your energy expenditure, P. Moore, 18 May 2011, http://triathlete-europe.competitor.com/2011/05/18/how-to-calculate-your-energy-expenditure, accessed 22 March 2014.
[16] Walking and running energy efficiency, http://www.exrx.net/Aerobic/WalkCalExp.html, accessed 22 March 2014.
[17] TrainingPeaks power profiles for cyclists, A. Coggan, Bicycling.com, December 2012, http://www.bicycling.com/training-nutrition/training-fitness/trainingpeaks-power-profiles-cyclists, accessed 24 March 2014.
[18] Human powered aircraft, Wikipedia, 2 February 2014, http://en.wikipedia.org/wiki/Human-powered_aircraft, accessed 24 March 2014.
[19] Askew, G.N. & D.J. Ellerby, 2007: The mechanical power requirements of avian flight, Biology Letters, 3(7), 445-448, 22 August 2007, http://www.ncbi.nlm.nih.gov/pmc/articles/PMC2390671/, accessed 24 March 2014.
[20] Lock, R.J. Et al., 2013, Multi-modal locomotion from animal to application, Bioinspiration and Biomimetics, 9, 011001, doi: 10.1088/1748-3182/9/1/011001, 16 December 2013, http://m.iopscience.iop.org/1748-3190/9/1/011001/article, accessed 24 March 2014.
[21] TNT equivalent, Wikipedia, 14 March 2014, http://en.wikipedia.org/wiki/TNT_equivalent, accessed 24 March 2014.
[22] Thermonuclear weapon, Wikipedia, 22 February 2014, http://en.wikipedia.org/wiki/Thermonuclear_weapon, accessed 24 March 2014.
[23] Temperature of a nuclear explosion, Hypertextbook.com, S. Fung, 1999, http://hypertextbook.com/facts/1999/SimonFung.shtml, accessed 24 March 2014.
[24] Nuclear weapon yield, Wikipedia, 19 February 2014, http://en.wikipedia.org/wiki/Nuclear_weapon_yield, accessed 24 March 2014.
[25] Effects of a nuclear explosion, Wikipedia, 17 March 2014, http://en.wikipedia.org/wiki/Effects_of_nuclear_explosions, accessed 24

Measuring the World, by John Austin

March 2014.
[26] Alvarez, L., W. Alvarez, F. Asaro and H.V. Michel, 1980: Extraterrestrial cause for the Cretaceous-Tertiary extinction, Science, 209, 1095-1108, accessed 24 March 2014.
[27] How much energy does a hurricane release?, C. Landsea, NOAA Hurricane Research Division, 2014, http://www.aoml.noaa.gov/hrd/tcfaq/D7.html, accessed 24 March 2014.
[28] Orders of magnitude (energy), Wikipedia, 26 February 2014, http://en.wikipedia.org/wiki/Orders_of_magnitude_(energy), accessed 24 March 2014.
[29] Aviation and the global atmosphere, Intergovernmental Panel on Climate Change, Special Reports on Climate, 1999, Summary for policy makers, http://www.grida.no/publications/other/ipcc_sr/?src=/climate/ipcc/aviation/006.htm#spm41, accessed 25 March 2014.
[30] Air travel, Climate Change Connection, 22 November 2010, http://www.climatechangeconnection.org/solutions/Airtravel.htm, accessed 25 March 2014.
[31] How much fuel does an international plane use for a trip?, HowStuffWorks.com, 1 April 2000. http://science.howstuffworks.com/transport/flight/modern/question192.htm, accessed 25 March 2014.
[32] Jet engine, Wikipedia, 16 March 2014, http://en.wikipedia.org/wiki/Jet_engine, accessed 25 March 2014.
[33] Journey through a jet engine, Rolls-Royce, http://www.rolls-royce.com/interactive_games/journey02/flash.html, accessed 25 March 2014.
[34] The GE90 family, GE Aviation, 2012, http://www.geaviation.com/engines/commercial/ge90/, accessed 25 March 2014.
[35] Rocket engine, Wikipedia, 21 March 2014, http://en.wikipedia.org/wiki/Rocket_engine, accessed 25 March 2014.
[36] Saturn V, Wikipedia, 17 March 2014, http://en.wikipedia.org/wiki/Saturn_V, accessed 25 March 2014.
[37] Space shuttle, Wikipedia, 24 March 2014, http://en.wikipedia.org/wiki/Space_Shuttle, accessed 25 March 2014.
[38] Fireworks, Wikipedia, 25 March 2014, http://en.wikipedia.org/wiki/Fireworks, accessed 25 March 2014.

[39] Components of natural gas, Enbridge Gas, 2014, https://www.enbridgegas.com/gas-safety/about-natural-gas/components-natural-gas.aspx, accessed 25 March 2014.
[40] Heat of Combustion, Wikipedia, http://en.wikipedia.org/wiki/Heat_of_combustion, accessed 25 March 2014.
[41] Hydrogen vehicle, Wikipedia, 12 March 2014, http://en.wikipedia.org/wiki/Hydrogen_vehicle, accessed 25 March 2014.
[42] Farrell, A.E., et al., 2006: Ethanol can contribute to energy and environmental goals, Science, 311, 506-508, doi: 10.1126/science.1121416, 27 January 2006, http://www.sciencemag.org/content/311/5760/506.short, accessed 25 March 2014.
[43] Simple machine, Wikipedia, 20 March 2014, http://en.wikipedia.org/wiki/Simple_machine, accessed 25 March 2014.
[44] Archimedes quotes, Brainyquote.com, http://www.brainyquote.com/quotes/authors/a/archimedes.html, accessed 25 March 2014.
[45] Wheel and axle, Wikipedia, 13 March 2014, http://en.wikipedia.org/wiki/Wheel_and_axle, accessed 25 March 2014.
[46] Pulley, Wikipedia, 16 March, 2014, http://en.wikipedia.org/wiki/Pulley, accessed 25 March 2014.

12. Electrical Current and Electrical Power
Current: the movement of electrical charges in a conductor

12.1 Introduction

Numerous devices rely on electrical power from batteries or directly from the electricity grid[1]. Electricity is now also an important commodity with associated international trading[2]. Electrical energy and power can be developed from the starting point of electrical current, which is a measure of the charges flowing along a wire, for example. The French physicist Ampere provided the mathematical theory to explain how electrical currents exerted forces upon themselves, and so the definition of current, the Ampere, bears his name, as described in § 12.2. From the Ampere, a number of other units are defined, which leads us back to the joule as the unit of energy. Moreover there is global agreement in the use of the metric system, for current and the vast majority of electrical units: there is no equivalent fps current. Further electrical properties capacitance and inductance are described in § 12.3 – 12.4. Although these properties are less known than charge and current flow, they are important properties of materials which influence a wide range of consumer products. In § 12.5 a brief introduction is given on how these properties are measured.

Electricity is one of a range of physical phenomena which are encompassed by the general term electromagnetism. It includes magnetism (§ 12.6) which is an important property of matter that is often exploited to our advantage, for example in nuclear magnetic resonance machines in hospitals (§ 12.7), or simply in the use of a compass for navigation.

Electromagnetic phenomena – electricity and magnetism – are linked by an important set of equations called Maxwell's equations. These equations were first established in 1861 and 1862[3,4] and relate changes and variations in the electric field with changes and variations in the magnetic field. In their modern form they represent an elegant way of combining the known electricity and magnetism laws into a single set of simple differential equations (which nonetheless require advance advanced mathematics to understand!). The equations were crucial in the development of the theory of special relativity[5]. For example an observer moving through a magnetic field doesn't just observe a magnetic field, but a combination of an electric and magnetic fields, in the same way that a moving charge experience both a magnetic and electric field, according to Maxwell's equations.

Lightning is one of several natural sources of electricity, including that wielded by the animal kingdom, as described in § 12.8. People of course have devoted a large amount of resources into generation of various kinds described in § 12.9 – 12.12. The issue of energy efficiency is raised in § 12.13 – 12.14.

12.2 Electrical Current, Voltage and Resistance

Electrical current is the movement of electrical charges in a conductor. One ampere (A) is the current which when flowing in two parallel wires a metre apart in a vacuum, produces an opposing force of 2×10^{-7} newtons per metre length[6]. The definition uses the force law established by Ampere. The net charge in coulombs, C, is the product of the current and the time that it acts for.

Measuring the World, by John Austin

The next electrical unit that is needed is the volt (V), which is a derived unit in the SI system. One volt is the electric potential between two points when a current of one amp dissipates one watt of power[7]. The power unit has been defined previously as a mechanical power (Chapter 11), so the volt is a *derived* unit. The product of the current, *I* (amps) and voltage, *V* (volts), *IV* is the power consumed in watts (W). When multiplied by the time in seconds over which the power operates, we obtain the energy used in joules.

The electrical resistance of a material is a measure of how much it opposes an electrical current. It is measured in ohms[8], Ω. The definition of the ohm has changed over the decades and is now the electrical resistance between two points when a constant potential difference of one volt between these points produces a current of one amp. Given a resistance *R* (ohms) and current *I* (amps) the electromotive force (emf) or voltage, *V*, to drive that current is $V = IR$, which is known as Ohm's law. So the ohm is now also a unit derived from the volt, which in turn is derived from the amp and watt.

Until relatively recently (1948), the International Standard Ω was the electrical resistance of a mercury column 106.3 cm long and 1 mm^2 cross section at 0 °C[8], thus providing a physical standard just like the metre bar, but this has now been abandoned in favour of universal units and the definitions work via the ampere.

Materials that have a high resistance have poor conductivity and the converse. It is therefore sometimes convenient to have a measure of conductance, the opposite of resistance. The units for this are Siemens (S) = Ω^{-1}, The unit is also sometimes referred to as mho (name spelled backwards, symbol upside down $\mho$). Conductivity is high in the presence of free electrons which carry charge along a wire for example. Aluminium (Al), copper (Cu), silver (Ag) and gold (Au) are all good conductors, with more than one free electron per atom. Lead (Pb) and Tin (Sn) are poor conductors with less than one free electron per atom.

12.3 Capacitance

A capacitor is a device for storing electrical charge. In its simplest form, it consists of two plates placed a distance apart with a voltage connected across them. Charges are induced on the plates, and the charge increases with the voltage. A measure of the storage capacity is the capacitance given by $C = Q/V$ where Q is the total charge and V the voltage. When the charge is measured in coulombs and the voltage in volts (the SI units for charge and emf respectively), the capacitance is measured in the SI unit of capacitance, farads (F), after Michael Faraday (Appendix C). However, a farad is a very large charge, because the coulomb on which it is based is also a very large charge to be stored in a single location. Capacitors[10] are often measured in µF or pF. The devices are especially useful in electronic circuits, for example, where a rapid discharge may be needed.

12.4 Inductance

An important step in the understanding of electromagnetism was the discovery of electrical inductance[11] independently by Faraday and Henry. The discovery led eventually to the development of electric motors and electromagnetic radiation (§ 5.6). Essentially, a changing current induces a voltage which opposes the change in current in an electric circuit. The magnitude of the voltage depends on the rate of change of current: $V = -LdI/dt$. dI/dt is the rate of change of current, measured in $A\ s^{-1}$, and if the induced voltage V is measured in volts, the inductance, L, is measured in henries (H). The minus sign in the above equation is a mathematical way of indicating that the voltage induced opposes the current change. A transformer[12] exploits inductance to change the voltage of an electrical signal. This usually operates more efficiently with alternating current (a.c.) and is achieved by winding a wire around a metallic core (see diagram). The ratio of the voltages $V_S/V_P = N_S/N_P$ = the ratio of coils around the core, where subscript S is for the secondary (usually output) and subscript P is for the primary or input. Transformers are essential to change the voltage of the power generated from a power station to the lower voltage (220/240 V in Europe, 110 V in the USA) used in our homes.

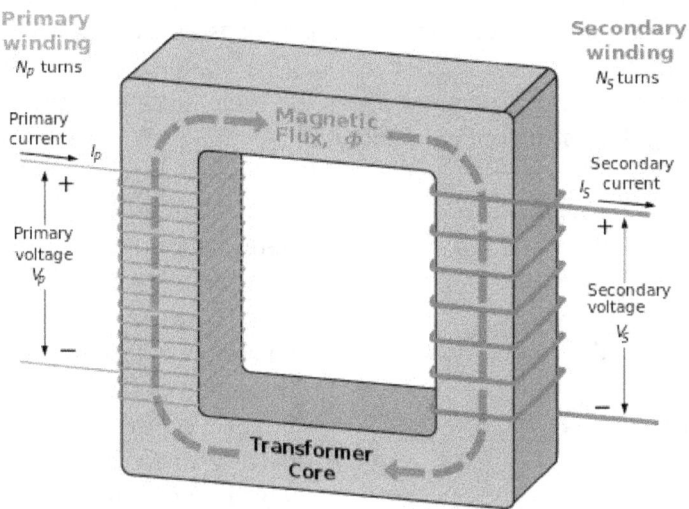

Idealised single-phase transformer showing the path of the magnetic flux through the core. Taken from Wikimedia[13]. The secondary voltage, V_S is given in terms of the primary voltage V_P by $V_S/V_P = L_S/L_P$.

12.5 Measuring Electrical Properties

The most reliable devices now available for measuring simple properties such as current, voltage and resistance, are digital meters[14]. For example, voltage can be compared to a standard within the instrument and an analogue to digital converter can be used to display the result. A digital multimeter can measure current, voltage and resistance. In an analogue meter a coil of wire is mounted inside a magnet. The coil rotates when a current passes through because of the induced magnetic field, and a needle attached to the coil turns to indicate the current. The instrument can be adjusted to measure the current through the standard internal resistor – it then measures voltage. Or the instrument can be modified to apply a standard voltage to an unknown resistor. The current passing then indicates resistance – making the instrument an ohmmeter.

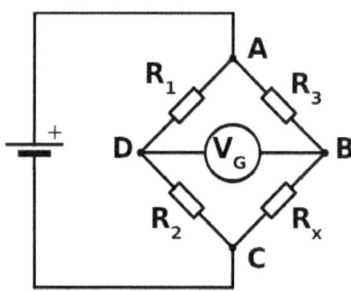

An alternative way of determining resistance is to use the instrument to detect a non-zero current. It is then known as a galvanometer. In the set up of the Wheatstone bridge[15], the apparatus is used to find the condition in which no current flows through the galvanometer. The measurement accuracy then depends not on the instrument accuracy but on its sensitivity. In other words, if zero current is detected how much does the set up need to change for the galvanometer to register a non-zero current?

In the Wheatstone bridge (above, from [16]), R_x is an unknown resistor, R_1, R_2 and R_3 are known resistances. R_1, R_2 and

R_3 are adjusted until the galvanometer (V_G) signal is zero. We then have $R_x/R_2 = R_3/R_1$ from which R_x can be determined. To allow a higher precision determination to be made, R_1 and R_3 are sometimes substituted by high resistance wire, and the exact ratio R_3/R_1 can be determined by sliding a pointer along the wire.

Utility meters[17] in homes measure electricity consumed, or the flow of water or gas. They contain electric motors that turn when a current or flow of water or gas occurs, and the meters use a system of gears. In analogue meters, the gears are usually oppositely directed going from small-sized units on the right, in which the pointers often move clearly, like the second hand of a clock. On the left, the pointers move much more slowly , like the hour hand of a clock. However, digital meters are generally more common now.

12.6 Magnetism

Magnets always appear in the form of a dipole: with a 'North' and 'South' pole. On a bar magnet, the poles are often marked N or S and the North pole of the magnet points towards magnetic north on the Earth. If you were to cut the bar magnet in half, you would simply have two bar magnets, each with N and S poles. The geomagnetic pole, where a magnet points downwards with no horizontal direction does not coincide with the Earth's axis rotational pole. Instead, the geomagnetic pole varies in its position over a period of a century or more[18]. In 2007, the North magnetic pole was at 84 °N, 121 °W and the South magnetic pole was at 65 °S, 138 °E. As long ago as 1831 the North magnetic pole was at only 70 °N. Maxwell's equations allow for the existence of electric monopoles (charges) but not for magnetic monopoles, although the equations can easily be modified to allow for them[19]. However, despite searches for their effects, magnetic monopoles have never been discovered anywhere in the universe.

The SI unit of magnetic field is the tesla (T), and in the cgs system, the gauss (G) is used. Both units are named after scientists who did important work in electromagnetism. One tesla is a very strong field, and 1 T = 10^4 G, compared with Earth's magnet field of about ½ G. It arises from the electrical currents in the Earth's rotating molten core. In observations of the Earth and planets, magnetic measurements are made as it provides amongst other things, valuable information about the nature of the planet's core[20]. The total magnetic force on an object in a magnetic field is proportional to the magnetic flux intercepted by the object. The force is the product of the field strength in T and the area intercepted in m^2. This is measure in webers (Wb), where 1 Wb = 1 T m^2. The magnetic force is the force that we feel when we place an object in a magnetic field. It could be another magnet, or another metallic object.

12.7 Societal Uses of Magnetic Fields

Magnetism is used frequently in modern society. As we have seen above, they are frequently used in instruments to measure electricity and in compasses to determine the direction of magnetic north. For navigation purposes, though, compasses are now being superseded by GPS tracking. Magnetic fields are also exploited in hospitals in magnetic resonance imaging (MRI) scanners, and in transport systems, magnetic levitation (maglev) trains are becoming more common.

Magnetic fields polarise hydrogen atoms. In MRI scanners[21], the tissue is exposed to a strong oscillating magnetic field of about one T. The magnetic field aligns the protons in the tissue, essentially the hydrogen atoms, and the positions of those atoms oscillate. The oscillations emit radio waves which can be detected for imaging, allowing active and passive cells to be identified in the body, especially in the brain. The magnets are

Measuring the World, by John Austin

protected in these machines since magnetic objects can acquire a large force in their vicinity. The magnets need to be cooled with liquid helium to be superconducting. Hospitals, then, provide an excellent example of the adoption of modern physics understanding in both low temperature science and high magnetic field theory.

Magnets in hospital machines need to be protected from damage by stray metallic objects, such as coins and wristwatches. So people nearby need to remove all metallic objects from their person or pockets. It is somewhat complicated but the force on a metallic coin, for example, would be the force on two magnetic dipoles. One of the dipoles (magnets) is the MRI scanner itself, and the other is the coin, which may have been exposed to magnetic fields and be magnetised at say a small fraction of the MRI scanner's magnetic field. The force on the dipoles is[22]

$$3\mu_o m_1 m_2/(2\pi x^4)$$

In this expression, x is The distance between the two magnets (the distance between the machine and the person carrying the offending object), m_1 and m_2 are the magnetic moments of the two magnets and μ_o is the magnetic constant ($4\pi \times 10^{-7}$ in SI units H m^{-1}). This would be a difficult calculation in practice, but to calculate just an order of magnitude estimate of a force on a magnetised coin, we can take the MRI machine as having a field, B, of 1 T at a distance r from the machine and a magnetic moment $m_1 \approx B \times 4\pi r^3/\mu_o$ for a very approximate value, so for $r = 1$ m, $m_1 \approx B/1\times 10^{-7} \approx 1\times 10^7$. The coin becomes magnetised due to the strength of the MRI magnet a metre or more away, so the field would be much lower, say 1% of the MRI magnet, 1×10^{-2} T. Hence $m_2 \approx B \times 1\times 10^{-6}/1\times 10^{-7} \approx 1\times 10^{-2} \times 1\times 10^7 \times 1\times 10^{-6} \approx 0.1$. Putting in all these values into the dipole force, we obtain a force on the coin of 0.6 N at a distance of 1 m from the scanner. This does not sound very much, but it is roughly 10 times the weight of the coin, assuming it has a mass of 5 g, so it needs to be held quite firmly. The above estimates were difficult enough to complete in the SI system. We were able to exploit the fact that if fields and field densities were kept in the SI units, then the resulting answer would be in the SI unit for force (N). The only way that I can see of completing these estimates in the fps system would be to do the calculations in metric first and then convert the final answer of force into a force or acceleration in fps. If you tried it this way, did you miss any stray *g* factors?

Continuing on the health theme, mobile phones have been implicated in some brain cancers, but this remains controversial as a physical process is yet to be properly identified[23]. For example, the magnetic field generated by phone use is only of order 1-6 µT[24], considerably smaller than the Earth's magnetic field. Nonetheless, the World Health Organization is concerned about the risk to people of electrical objects and have issued guidance concerning the impact of electromagnetic fields in general[25]. Many electrical devices emit magnetic fields comparable or larger than mobile phones, but the latter are unique in being held so close to the head for long durations in some cases, hence the concern and need for more research.

Another area where strong magnetic fields have provided an important benefit to society is in the development of maglev trains[26]. The trains carry magnets on their base and are levitated about 1 cm above the guide rails by magnetic forces from the magnetic field induced by electric currents in the rails. By alternating the polarity of the magnetic signal in the rails, the trains are accelerated by a sequence of attracting and repelling forces. The trains are effectively low-flying aircraft and can exceed speeds of 500 km h^{-1}. I experienced one such journey from Shanghai airport (China) to the city. It was a very smooth ride, smoother than anything I've been on before or since. Acceleration is much faster than a normal train, because the force which accelerates a normal train is limited by the friction on the rails, which is necessarily low. However, I suffer a bit from travel sickness, and I found it a bit disorienting when the train tilted to negotiate the curves on the route. It took only 8 *min* for 40 *km* and it was worth spending a few extra dollars to cut the journey time. Unfortunately the trains are quite expensive to run and have not taken off (sic!) in the world at large.

12.8 Natural Electricity

In 1752 Benjamin Franklin demonstrated that lightning is an electrical discharge[27]. The story indicates that he flew a kite with an electrical conductor attached, into a thunder cloud. He would therefore have used his body as an electricity meter. Ouch! This is not recommended at home. In fact it seems that this is one of numerous apocryphal stories surrounding scientists and their work and may not be strictly true. In air, the electrical current from a lightning discharge is typically about 30 kA with 500 MJ of energy[28]. During the discharge, air becomes heated to a temperature of up to about 50,000 °C and the rapidly producing pressure wave moves away from the source faster than the speed of sound, causing thunder.

To avoid lightning, stay indoors or in an enclosed vehicle with the windows up, but don't touch any metal. It is not safe near tall objects such as a tree or on open metal objects such as a bicycle. Be the smallest object around if it is unavoidable to stay out during a thunderstorm. A lightning rod attached to a building attracts lightning preferentially to it, conducting the electricity harmlessly to earth.

While hopefully, we don't make a habit of it, electrical energy in the form of a shock can feel like a force on our bodies. In fact the severity of the shock depends on the current. For example, skin resistance is about 0.1 MΩ when dry, but reduces to about 1 kΩ when wet[29]. Hence a voltage of 2000 V will generate a current of 20 mA in the first case, but 2 A for wet skin. A current of 30 mA for a fraction of a second is sufficient to induce fibrillation and cardiac arrest[29].

There are many examples of the use of electricity by the animal kingdom, the most widely known of which are stingrays, eels and catfish[30]. The electric ray delivers a shock to its prey or as protection and have been known to kill people. For example, the "crocodile hunter", Steve Irwin was killed by a stingray probably as a defence measure and Irwin died at the spot[31]. The electric eel applies a voltage of about 10 V for orientation and locating prey, and a powerful 650 - 700 V burst for predation and defence[30]. Catfish use electricity in a similar way as electric eels. While the power available to these fish is sufficient to kill a person in principle, unlike in the tragic stingray case described above, I could find no confirmation. By contrast, human deaths and injuries from jellyfish are well documented[32] and much more common, but the mechanism operates via venom[33], not electricity.

12.9 Power Stations

Power stations typically work on the principle of generating heat, which is then used to raise steam to drive turbines. It is an inherently inefficient method. One US government site[34] defines the efficiency as "dividing the equivalent BTU content of a kW h of electricity (which is 3412 BTU) by the heat rate", illustrating the mixed up units in existence in the USA. For US sources, fossil fuels work out at about 31-33% efficient for coal and petroleum, but as high as 42% for natural gas. Nuclear power is just as inefficient as coal and petroleum at 32%. Other information[35] suggests that the efficiencies may be 5% higher in modern designed plants. Heat is usually generated by burning something, which has the added problem of producing CO_2 pollution. If coal is burnt, there is often also mercury products emitted into the atmosphere. Even more pollution is associated with sulphur-containing coal. "Clean coal technologies"[36] have been invented to address some of the environmental issues regarding the burning of coal, but the term is very vague. In the past, CO_2 and mercury have still been emitted, but even those are now being addressed. Unfortunately, the world's largest polluter, the USA, is in denial mode regarding atmospheric emissions. In the 2012 presidential campaign, the Republican challenger, Mitt Romney, had an energy 'plan' which relied heavily on fossil fuel production[37]. He maintained that the scientific consensus on the extent of atmospheric warming and the degree to which humans have contributed is still lacking. Neither argument is really true, as copious reports from the Intergovernmental Panel on Climate Change[38] have testified. Under the Democratic president's watch meanwhile, coal production and consumption in the USA have continued to increase[39].

A typical power station output is about 1 GW, but can be larger, for example 4 GW for Drax[40], in the UK. US coal fired stations had a total capacity of about 278 GW in 2000 and increased to 339 GW in 2009[41]. The actual electrical production in 2000 was 20% less at 224 GW. Annual energy production for the USA as a whole is for some strange reason measured in quads = 10^{15} *BTU* (i.e. 1 quadrillion *BTU*). 300 GW for a year is 300x10^9 x 365 x 86400 J = 9.47x10^{18}J = 9.47 EJ. Since 1 BTU = 1054 J (§ 11.3), 300 GW converts to 8.98 *quads* per year. Power station output is alternating current with a polarity changing from positive to negative in a sine wave with a frequency 50 or 60 Hz. Since power loss in transmission lines is proportional to current squared, $IV = I^2R$, transmission losses are minimised for minimum current and maximum voltage. Power is therefore generated at thousands of volts (typically 11 – 33 kV[42]). There is a trade-off, though, in that with higher voltages, insulation needs to be made more effective. Power is then efficiently transformed to low voltage (220/240 V in Europe, 110 V in USA) for use in the home, with only a small loss of energy. In principle, 240 V is slightly safer as it leads to lower currents. This could be important for people receiving accidental electric shocks, as an equivalent energy would be delivered at lower current in Europe.

Worldwide in 2008, total electrical power production was 2.3 TW[43]. In the last 10 years, most types of electricity production have grown, especially renewables and fossil fuel, while production from nuclear has remained about the same. Proportions vary substantially from one country to the next depending on their mix of natural resources. Worldwide, in 2008, Coal accounted for 41% of electrical energy production, gas 21%, oil 5%, nuclear 13% and renewables 16%. There is an unspecified 3%, and presumably arises from the burning of something.

12.10 Nuclear Power: Fission

Nuclear power stations do not produce CO_2, so do not contribute to global warming and for this reason there has been some resurgence in their commissioning. However, environmental issues remain, regarding the disposal of radioactive waste and cost over-runs[44]. Also the Tsunami which destroyed the Dai Ichi nuclear power station in Fukushima, Japan in 2011[45] has brought back more environmental concerns. Japan's power production was immediately affected, and after enforced reductions in power supply, many people willingly changed their domestic activities to reduce their electricity bills[46]. Japan itself has declared that it will be closing its nuclear power plants by 2040, but that is a long way away for a politician and it didn't take them long to reconsider[47], amid renewed concerns about lowering standard of living.

Measuring the World, by John Austin

Nuclear power works on the principle of nuclear fission: heavy atoms are broken down into smaller fragments with the release of energy[48]. The mass of the smaller fragments is less in total than the mass of the initial atoms, and by the conservation of mass and energy, the difference appears as energy according to Einstein's famous equation $E = mc^2$. The nuclear fragments are typically radioactive requiring careful storage for centuries. Despite these rather large environmental problems, huge amounts of energy are realisable. For example, just 1 g of matter converted to energy produces $1 \times 10^{-3} \times (3 \times 10^8)^2 = 9 \times 10^{13}$ J, enough to generate 1 GW of power for a day. Nuclear power stations typically use U-235 which is only present in natural uranium at a concentration of 0.72%[49]. Some specialised reactors used enriched uranium, in which the uranium contains over 5% U-235. Fast reactors exist which can convert waste to energy. That waste includes plutonium and americium etc. One kilogramme of uranium provides as much energy as 1500 t of coal. During the fission process, heat is produced which needs to be kept under control by inserting a 'moderator' into the fuel, which is kept in specially prepared fuel rods. Fission also produces excess neutrons, which are absorbed outside the uranium core. This generates heat which is used to drive turbines and then electricity in the usual way.

12.11 Nuclear Power: Fusion

Fusion reactors, in which atomic nuclei are fused together to generate energy, are still in the research stage. Once commercial stations are available, the environmental cost should be solved, as the waste products quickly lose any radioactivity. To fuse two atoms is scientifically and technologically more challenging than fusion, because fusion works against the forces which repel the nuclei[50]. If, however, the nuclei can be made to move at very high velocities, the repulsion can be overcome.

Measuring the World, by John Austin

Atomic velocities translate directly to temperature, so essentially the question is how to raise their temperature in a controlled fashion. Fusion occurs in the core of the sun at temperatures of 15 million K[51], so the problem is how to get to these temperatures in a controlled environment, as opposed to an uncontrolled environment in an H-bomb. At very high temperatures, atoms no longer remain intact: the electrons surrounding the nucleus get stripped off, leaving the positively charged nuclei. Such a material is known as a plasma, the 4th state of matter after solid, liquid and gas[52]. While not as generally well-known as the other three states of matter, plasmas still occur frequently in nature and in industrial applications. For example, plasmas are present as the glowing 'gas' inside neon signs and sodium street lamps. In nature, plasmas are produced in the air in a lightning discharge and in the aurora visible sometimes in the high latitude sky. Much of interstellar space at the stars themselves consist of plasma, with only a very small proportion of the volume and mass of the universe contained in the solid, liquid and gas phases. We just happen to have a preference for them!

If a hot plasma touches any solid material, then it will immediately lose its heat to the solid. It is therefore a considerable challenge to contain a hot plasma for fusion reaction purposes. In practice, reactors are generally doughnut shaped (tokomaks) with extremely strong magnetic fields produced around the doughnut[53]. This helps to confine the plasma to the interior of the vessel. Research has reached the stage where the fusion reactions are close to 'ignition'. In other words, almost as much energy is produced by fusion as is required to heat the plasma. Performance generally increases with the size of the reaction vessel. The next phase for fusion research is the International Thermonuclear Reactor (ITER)[54,55] which is about 30 m tall overall, and contains a reaction vessel of 840 m^3, the largest of any machine built. First plasma is expected at the end of 2020, and maximum performance is expected by 2027. Temperatures are planned to reach 150 MK and the machine is designed to produce 500 MW of power for up to 1000 s at a time, 10 times more than the energy required to heat the plasma.

In the sun, the hot plasma is confined by gravity, which is a significant force for an object of its size. Moreover, the mean density of the sun is 1400 kg m^{-3} compared with ITER density of 0.5 g of plasma in 840 m^3, or a density 2 billion times lower than the sun. This explains why it is so difficult to generate energy this way!

Another approach to creating laboratory fusion is by using lasers to focus their energy onto a suitable fuel pellet[56]. This is being tried at Lawrence Livermore National Laboratory in the USA[57] where 192 lasers concentrate their power into a pellet containing deuterium and tritium, isotopes of hydrogen and the nuclei most likely to fuse. The lasers provide peak energy of about 2 MJ and as recently as September and November of 2013 showed an energy gain[58], with 10 kJ to the pellet, and 15 kJ of fusion energy released. While break even is a significant achievement, plasma ignition, in which the reaction in the plasma becomes self-sustaining has not yet been reached.

'Cold fusion' has been investigated. In earlier experiments fusion energy was purported to have been produced at room temperature, but the process cannot be identified and is no longer considered viable[53].

12.12 Renewable Power

Renewable power is essentially electrical power generated by a system with no explicit fuel source[59], and represents about 17% of total energy consumed worldwide. It has the advantage of not producing net greenhouse gases, but it is not without its environmental cost. The absence of an explicit fuel source does not of course contradict the law of conservation of mass and energy. The fuel for renewable energy comes typically from the sun (solar, wind, wave power). Even hydro power comes ultimately from the sun, as the water driving the turbines has participated in a cycle of evaporation and condensation driven by the sun. Geothermal energy comes directly from the Earth.

Amongst the renewable power systems currently in use, wind power has a prominent role in the UK. Wind turbines typically extract about 40% of wind energy. The maximum is thought to be 59% [60]. Geothermal heating is the main physical process behind geysers[61]. These are particularly well-known in Iceland, 110 km from Reykjavik, and also in New Zealand, and Yellowstone National Park (USA). Cold water seeps down from nearby lakes and rivers, filling channels in the rock. Well below the surface, the temperature is as much as 300 °C. The cold water is heated by the rocks, but remains liquid under high pressure. Eventually, the temperature of the water increases enough for steam to be released. This occurs in a burst, simultaneously lifting the water column at lower depths. The process accelerates up through the column with more water turning to steam. Overall, the burst of steam occurs irregularly, but can be made regular (to avoid disappointing the tourists) by adding soap powder! This process can be exploited to heat the home, but in a more controlled manner[62]. By passing a pipe containing cold water to below the surface and around in a loop, the rising pipe can contain hot water by conduction from the surrounding ground. In summer the sub-surface temperature could be lower than above the ground, so the same system can be used to cool the home, when combined with fans.

12.13 Off-Peak Power Use

Nuclear and fossil fuel power stations work most efficiently when they run continuously with only a small variation in power output. However, power demand tends to occur during certain peak hours. Markets price electricity to shift demand from those peak times to off-peak times. Electricity is an expensive way to heat homes, for example, as it transfers high-grade energy (which can be used for anything) to low-grade energy (namely heat). It is better to burn natural gas locally to warm the home and cook than to use electricity. Storage heaters[63] reduce the cost difference by taking advantage of low-priced off-peak electricity to heat a ceramic block or clay bricks within the heaters. The block is thermally insulated until the heat is needed. The heat of the block is then released slowly, as needed, without further electricity consumption. To reduce electricity bills, it helps to have a similar pattern of daily heating. Heating bills can, with this strategy, be more competitive. Although still more expensive than heating with natural gas, storage heaters can be used where natural gas is not available. Overall, though, storage heaters are not very environmentally friendly.

12.14 Energy Efficiency in the Home

Having a more energy efficient home, which can be done in a number of ways[64], will reduce energy bills and reduce climate change. One of the simplest measures to take is to replace all incandescent light bulbs. Many countries already have at least partial bans in place[65]. The USA and Canada as usual are doing their bit for the global environment by doing nothing about them. The bulbs are grossly inefficient for light production, as over 95% of the energy consumed goes straight to heat. Incandescent bulbs were invented by Edison in 1879 and have had a long history[66], but their replacement can save many kW h during their lifetime. The energy saving by switching to compact fluorescent is readily explained on the packaging. Many people find the light from these bulbs too white or blue-white. This is really the human tendency to dislike change, as incandescent bulbs actually give off an ugly yellow-brown light that some people have become accustomed to. For example, as an experiment take a digital photograph of a colourful object, particularly with reds and yellows but preferably the full spectrum of colours, under an incandescent light (no flash) Then do the same in daylight or with a flash. By comparing the colours you see what you could be missing under incandescent lighting. In any case more recent development of light emitting diode (LED) technology will likely completely transform lighting within a few decades. In particular LEDs can be adjusted for colour temperature (and hence can be made more orange, or ugly yellow-brown as the mood dictates). They can also be used in dimmer switches without loss of efficiency. LEDs are discussed in more detail in § 14.5.

References

[1] Electricity, Wikipedia, 26 March 2014,
http://en.wikipedia.org/wiki/Electricity, accessed 26 March 2014.
[2] Electricity market, Wikipedia, 24 March 2014,
http://en.wikipedia.org/wiki/Electricity_market, accessed 26 March 2014.
[3] Maxwell's equations, R. Nave, Hyperphysics, Electricity and magnetism, Georgia state University, http://hyperphysics.phy-astr.gsu.edu/hbase/electric/maxeq.html, accessed 26 March 2014.
[4] Maxwell's equations, Wikipedia, 25 March 2014,
http://en.wikipedia.org/wiki/Maxwell's_equations, accessed 26 March 2014.
[5] Hall, G., 2008: Maxwell's electromagnetic theory and special relativity, Phil. Trans. R. Soc. A 28 May 2008, 366, 1849-1860, doi: 10.1098/rsta.2007.2192,
http://rsta.royalsocietypublishing.org/content/366/1871/1849.full,

accessed 26 March 2014.
[6] Ampere's force law, Wikipedia, 9 March 2014, http://en.wikipedia.org/wiki/Amp%C3%A8re%27s_force_law, accessed 26 March 2014.
[7] Volt, Wikipedia, 18 March 2014, http://en.wikipedia.org/wiki/Volt, accessed 26 March 2014.
[8] Ohm, Wikipedia, 3 March 2014, http://en.wikipedia.org/wiki/Ohm, accessed 26 March 2014.
[9] Electrical conductance, Wikipedia, 25 March 2014, http://en.wikipedia.org/wiki/Electrical_conductance, accessed 26 March 2014.
[10] Capacitor, Wikipedia, 26 March 2014, http://en.wikipedia.org/wiki/Capacitor, accessed 26 March 2014.
[11] Inductance, Wikipedia, 20 March 2014, http://en.wikipedia.org/wiki/Inductance, accessed 26 March 2014.
[12] Transformer, Wikipedia, 24 March 2014, http://en.wikipedia.org/wiki/Transformer, accessed 26 March 2014.
[13] Daniels, A., 1976: Introduction to electrical Machines, Macmillan publishers, ISBN 0-333-19627-9, 1 January 2006, http://en.wikipedia.org/wiki/File:Transformer3d_col3.svg, accessed 26 March 2014.
[14] Multimeter, Wikipedia, 6 March 2014, http://en.wikipedia.org/wiki/Multimeter, accessed 26 March 2014.
[15] Wheatstone bridge, Wikipedia, 14 February 2014, http://en.wikipedia.org/wiki/Wheatstone_bridge, accessed 26 March 2014.
[16] Wheatstone bridge circuit diagram, 9 October 2007, Author: Rhdv, http://en.wikipedia.org/wiki/File:Wheatstonebridge.svg, accessed 26 March 2014.
[17] Electricity meter, Wikipedia, 25 March 2014, http://en.wikipedia.org/wiki/Electricity_meter, accessed 26 March 2014.
[18] North magnetic pole, Wikipedia, 15 March 2014, http://en.wikipedia.org/wiki/North_Magnetic_Pole, accessed 28 March 2014.
[19] Magnetic monopole, Wikipedia, 13 March 2014, http://en.wikipedia.org/wiki/Magnetic_monopole, accessed 28 March 2014.

[20] Earth's core, magnetic field changing fast, K. Johnson, National Geographic News, National Geographic News, 30 June 2008, http://news.nationalgeographic.com/news/2008/06/080630-earth-core.html, accessed 28 March 2014.

[21] Magnetic resonance imaging, Wikipedia, 27 March, 2014, http://en.wikipedia.org/wiki/Magnetic_resonance_imaging, accessed 28 March 2014.

[22] Force between two magnetic dipoles, Physics Pages, http://physicspages.com/2013/06/25/force-between-two-magnetic-dipoles/, accessed 29 March 2014.

[23] Electromagnetic fields and public health: mobile phones, World health Organization, June 2011, http://www.who.int/mediacentre/factsheets/fs193/en/, accessed 29 March 2014.

[24] Magnetic field near a cellular phone, G. Elert, Hypertextbook, 2003, http://hypertextbook.com/facts/2003/VietTran.shtml, accessed 29 March 2014.

[25] What are electromagnetic fields?, World Health Organization, http://www.who.int/peh-emf/about/WhatisEMF/en/, accessed 29 March 2014.

[26] Maglev, Wikipedia, 27 March 2014, http://en.wikipedia.org/wiki/Maglev, accessed 29 March 2014.

[27] Did Ben Franklin really discover electricity when he flew his kite?, Curiosity.com, 2011, http://curiosity.discovery.com/question/ben-franklin-discover-electricity-kite, accessed 1 April 2014.

[28] Lightning, Wikipedia, 23 March 2014, http://en.wikipedia.org/wiki/Lightning, accessed 1 April 2014.

[29] Electric shock, Wikipedia, 29 March 2014, http://en.wikipedia.org/wiki/Electric_shock, accessed 1 April 2014.

[30] Electrical animals – shocking creatures, Electricshock.org, 2010, http://www.electricshock.org/electric-animals.html, accessed 1 April 2014.

[31] Steve Irwin, Wikipedia, 30 March 2014, http://en.wikipedia.org/wiki/Steve_Irwin, accessed 1 April 2014.

[32] Jellyfish stings in Australia, Wikipedia, 4 March 2014, http://en.wikipedia.org/wiki/Jellyfish_stings_in_Australia, accessed 2 April 2014.

Measuring the World, by John Austin

[33] Box jellyfish, Wikipedia, 1 April 2014, http://en.wikipedia.org/wiki/Box_jellyfish, accessed 2 April 2014.

[34] What is the efficiency of different types of power plants? Frequently asked questions, US energy information administration, 3 February 2014, http://www.eia.gov/tools/faqs/faq.cfm?id=107&t=3, accessed 2 April 2014.

[35] Compare the efficiency of different power plants, R. Scudder, Bright Hub engineering, 27 May 2010, http://www.brighthubengineering.com/power-plants/72369-compare-the-efficiency-of-different-power-plants/, accessed 2 April 2014.

[36] Clean coal technology, Wikipedia, 27 March 2014, http://en.wikipedia.org/wiki/Clean_coal_technology, accessed 2 April 2014.

[37] Mitt Romney's disastrous energy plan, Rolling Stone Politics, 14 September 2014, http://www.rollingstone.com/politics/news/mitt-romneys-disastrous-energy-plan-20120914, accessed 2 April 2014.

[38] Fifth assessment report, Intergovernmental Panel on Climate change, http://www.ipcc.ch/, accessed 2 April 2014.

[39] Short-term energy outlook, Independent Statistics and Analysis, US Energy Information Administration, 11 March 2011, http://www.eia.gov/forecasts/steo/report/coal.cfm, accessed 2 April 2014.

[40] Drax – facts and figures -power station, http://www.draxpower.presscentre.com/Facts-and-figures/Power-station-b89.aspx, accessed 2 April 2014.

[41] Coal power in the United states, Wikipedia, 28 March 2014, http://en.wikipedia.org/wiki/Coal_power_in_the_United_States, accessed 2 April 2014.

[42] Why generation voltage in power plant is low, Electrical questions guide, 15 March 2011, http://electricalquestionsguide.blogspot.co.uk/2011/03/why-generation-voltage-in-power-plant.html, accessed 2 April 2014.

[43] electricity generation, Wikipedia, 30 March 2014, http://en.wikipedia.org/wiki/Electricity_generation, accessed 2 April 2014.

[44] Nuclear power – the problems, Greenpeace, http://www.greenpeace.org.uk/nuclear/problems, accessed 2 April 2014.

[45] April 2011 Fukushima earthquake, Wikipedia, 18 March 2014, http://en.wikipedia.org/wiki/April_2011_Fukushima_earthquake, accessed 2 April 2014.
[46] In Japan, people get charged up about amping down, Wall Street Journal, 3 October 2012, http://online.wsj.com/news/articles/SB10000872396390443720204578003524193492696, accessed 2 April 2014.
[47] Japan says it will review plans to abandon nuclear power, BBC News Asia, 27 December 2012, http://www.bbc.co.uk/news/world-asia-20850416, accessed 2 April 2014.
[48] Nuclear fission, Wikipedia, 31 March 2014. http://en.wikipedia.org/wiki/Nuclear_fission, accessed 2 April 2014.
[49] Uranium, Wikipedia, 2 April 2014, http://en.wikipedia.org/wiki/Uranium, accessed 2 April 2014.
[50] Nuclear fusion, Wikipedia, 2 April 2014, http://en.wikipedia.org/wiki/Nuclear_fusion, accessed 3 April 2014.
[51] Solar core, Wikipedia, 7 March 2014, http://en.wikipedia.org/wiki/Solar_core, accessed 3 April 2014.
[52] Plasma (physics), Wikipedia, 2 April 2014, http://en.wikipedia.org/wiki/Plasma_(physics), accessed 3 April 2014.
[53] Fusion power, Wikipedia, 1 April 2014, http://en.wikipedia.org/wiki/Fusion_power, accessed 3 April 2014.
[54] ITER, The way to new energy, Official website, http://www.iter.org/, accessed 3 April 2014.
[55] ITER, Wikipedia, 9 March 2014, http://en.wikipedia.org/wiki/ITER, accessed 3 April 2014.
[56] Inertial confinement fusion, Wikipedia, http://en.wikipedia.org/wiki/Inertial_confinement_fusion, accessed 3 April 2014.
[57] National Ignition facility, Wikipedia, 1 April 2014, http://en.wikipedia.org/wiki/National_Ignition_Facility, accessed 3 April 2014.
[58] Hurricane, O.A. et al., Fuel gain exceeding unity in an inertially confined fusion explosion, Nature, 506, 343-348, 20 February 2014, doi:10.1038/nature13008, http://www.nature.com/nature/journal/v506/n7488/full/nature13008.html, accessed 3 April 2014.

[59] Renewable energy, Wikipedia, 1 April 2014,
http://en.wikipedia.org/wiki/Renewable_energy, accessed 3 April 2014.
[60] Wind turbine, Wikipedia, 2 April 2014,
http://en.wikipedia.org/wiki/Wind_turbine, accessed 3 April 2014.
[61] Geyser, Wikipedia, 7 March 2014,
http://en.wikipedia.org/wiki/Geyser, accessed 3 April 2014.
[62] How geothermal energy works, S. Watson, How Stuff Works,
http://science.howstuffworks.com/environmental/energy/geothermal-energy1.htm, accessed 3 April 2014.
[63] Storage heater, Wikipedia, 2 April 2014,
http://en.wikipedia.org/wiki/Storage_heater, 3 April 2014.
[64] Improve your home, Energy Saving Trust,
http://www.energysavingtrust.org.uk/Take-action/Improve-your-home
[65] Bans of incandescent light bulbs, Wikipedia, 25 March 2014,
http://en.wikipedia.org/wiki/Bans_of_incandescent_light_bulbs, 3 April 2014.
[66] Incandescent light bulb, Wikipedia, 30 March 2014,
http://en.wikipedia.org/wiki/Incandescent_light_bulb, 3 April 2014.

Measuring the World, by John Austin

Measuring the World, by John Austin

13. The Amount of Substance

Amount of substance = a measure of the number of atoms or molecules

13.1 Introduction

Chapter 7 explained the properties of materials based on their mass, a measure of inertia. An alternative measure of amount of substance relates to its chemical effects which are directly dependent on the numbers of atoms or molecules present. The French chemist Antoine Lavoisier and others[1] showed in the 18th century that equal masses of different substances did not react completely, and went on to form what might be considered the modern, quantitative understanding of chemistry. Rather, chemical reactions occur in fixed ratios. These ratios were used to set up a table of weights with each element represented. The weights are now known as 'atomic weights', and in the SI system is used to define the mole (mol):

1 mole of any element contains the same number of atoms as 12 g of Carbon-12 exactly.

Carbon-12 is the dominant isotope of carbon[2], which in nature is a mix of isotopes of mass 12, 13 and 14. Carbon-12 has 6 protons and 6 neutrons in the nucleus. In practice, the mean atomic weight of naturally occurring Carbon is close to 12, as the small amounts of C-13 and C-14 increase the atomic weight very little. The definition of the mole is very precise about the use of C-12 so that the standard is universally applicable.

Measuring the World, by John Austin

The atomic weight carbon, then is very close to 12 and likewise, the atomic weight of oxygen is 16. The molecular weight of a compound is the sum of its atomic weights. For example, oxygen gas, which is O_2 has a molecular weight of 32, twice 16. When the two react together, for example coal with its impurities sulphur, mercury etc. removed, 12 kg of carbon will react with 32 kg of oxygen to form carbon dioxide (CO_2), and the total mass of CO_2 will be 12 + 32 = 44 kg. In other words 1000 moles of carbon (C) and 1000 moles of oxygen (O_2) will react to form 1000 moles of CO_2.

In the 18th and 19th centuries, all known chemicals were explored with the idea of isolating the elements, those chemicals which could not be reduced to other elements, thus replacing the biblical fire, air earth and water concepts. Ratios in which the elements reacted enabled lists to be set up comparing the elemental masses. For example we have already seen how, if carbon is 12, then oxygen is 16. There was then a focus of scientific activity to understand the relationships between the elements, and this led to the development of the periodic table of the elements, described in § 13.2.

A major simplifying factor for gases is that 1 mole of most gases have the same volume. In fact gases do differ slightly, but if we ignore interactions between its molecules, a gas would follow exactly the kinetic theory of gases, and the gas is then known as an ideal gas. The ideal gas law and its predictions are described in § 13.3. With the definition of the mole established, an important aspect of scientific research has been to determine precisely how many atoms are contained in a mole. This is described in § 13.4, and is known as the Avogadro constant[3], after the Italian physicist. In recent years there has been more urgency in making high precision measurements of the Avogadro's constant, since under new ideas proposed for the SI units (Chapter 17), the definition of the mole will be changed. Of the practical application of the knowledge of molar quantities, calculations of gas volumes are described in § 13.3, while in § 13.5, concepts such as acidity use molar concepts.

These concepts are in principle convertible to fps units, which has its own set of constants used by certain adherents. For example, the mole is based on 12 g of C-12, so there are equivalent units based on 12 oz of C-12, but to my mind these appear contrived. So if one needs to calculate for example the density of a given volume of gas of known pressure and temperature, calculations are more easily done in the metric system and then converted back to fps. This is still a laborious and unnecessary calculation.

13.2 The Periodic Table and Atomic Weights

In the 19th century scientists were aware that certain elements had properties in common with others. This led to the development of the periodic table (*see figure, taken from TikZ [4]. The table has been corrected for minor typographical errors from the published version*). The version that we use today is due to the Russian physicist Mendeleev[5]. The table shows the elements ordered in terms of their properties and in terms of their atomic weights. The atomic weights are measured in daltons (Da or u), which is an 'accepted unit' in the SI system. For example, the atomic weight (the weight in g of one mole of substance) of C-12 is 12 Da, by definition of the size of the mole. In terms of the base SI units, 1 Da = 10^{-3} kg mol^{-1}. Each square in the table indicates an individual element with its name in full, and as the usual chemical symbol. It includes the atomic number (the number of protons in the nucleus) on the top left, and the atomic weight on the top right. The table includes all the naturally occurring elements, as well as those artificially produced. See the key in the table for an explanation of the colours (only in the ebook version).

Measuring the World, by John Austin

(Mendeleev's) Periodic Table of Chemical Elements via TikZ

Measuring the World, by John Austin

A detailed account of the periodic table is beyond the scope of the current work, but it is arguably one of the most impressive ways of collecting together chemical information. For a more complete explanation see [6].

Other periodic tables have been invented with the elements arranged differently[5], but Mendeleev's version has endured. Tables also exist showing the full range of nuclides (isotopes of different elements) that can occur[7] and these are considerably more detailed. Rather than showing chemical similarities, as with the standard periodic table, they show the connection of physical processes. The tables indicate the region of stability and the different types of radioactive decay for the unstable nuclides. In particular, there is a region of stability: a combination of protons and neutrons which keeps the nucleus stable, that is the nuclides are not radioactive. Beyond this combination, atoms become steadily more unstable until they can no longer form at all. For the layman it has perhaps at times seemed pointless to keep producing more and more atomic elements with heavier and heavier nuclei, but it has a useful aspect. That is, theory suggests that the nucleus has certain 'islands of stability' which occur for particular combinations of neutrons and protons[8]. Thus although heavy atoms are radioactive and decay in fractions of a second, a longer half-life than normal considering the size of the nucleus could be perceived as evidence for an 'island of stability'. Such regions of stability provide information about nuclear forces. Ordinarily, the nucleus is held together by the strong nuclear force, which tends to increase at short distances but decrease at larger distances. Eventually large numbers of nuclides cease to be stable, because of the force of repulsion between the protons. For low atomic weights, the numbers of neutrons and protons need to be about the same for stability, but as the atomic weight increases, more neutrons are required. Eventually the nucleons (protons and neutrons) move too far apart for the strong nuclear force to be effective, and the atom

Measuring the World, by John Austin

is unstable.

Atomic weights of naturally occurring elements vary between 1 for hydrogen (H) to 238 for uranium (U). Artificially produced atoms have reached 294 with an anticipated island of stability near atomic weight 300[8]. The atomic weight is approximately the same as the number of nucleons (protons + neutrons) in the nucleus. To determine the molecular weight, a small sample of material is vaporised, and passed into a mass spectrometer[9]. The atoms are charged and passed over a strong magnetic field. The curvature of their paths through the apparatus depends on the charge e and mass M. Knowing the charge and the curvature of the path reveals the value of M. M can be measured to very high precision (1 part in 10^6) in modern machines. It enables the ratio of different isotopes to be established.

High precision mass spectrometry is used in drug detection in conjunction with gas chromatography[10]. The sample is first added to the gas chromatograph, which separates the sample into its respective molecules. Essentially, higher atomic weight molecules take longer to pass through the gas chromatograph and can then be passed through the mass spectrometer. With a narrow range of masses entering the mass spectrometer, higher precision measurements can be made. Combined with a library of atomic weights, the mass spectrometer measurements can be used to identify individual molecules. With small molecules this is quite straightforward. For example if there is a need to separate two different chemicals which have about the same molecular weight. For example, to a first approximation, N_2 has molecular weight of 28, and so does C_2H_4. However, these are easily separated by precise measurements (28.0134 and 28.0532 respectively). As the molecular weight increases perhaps to the order of 1000 or more for many drugs, the precision of the mass spectrometer plays an important role in being able to identify chemicals.

Measuring the World, by John Austin

13.3 The Ideal Gas Law

The size of the mole and the chemical interactions of elements received a substantial boost from the understanding of gases, which turned out to be amenable to simple mathematics. Much of the empirical theory of gases, however, was put together by Boyle, Charles, Gay-Lussac and Avogadro, by the early 19th century in a serious of careful experiments[11]. Putting all the work together enables us to write down the gas equation

$$PV = mRT$$

This is known as the ideal gas equation or law. P is the pressure of the gas, and V its volume, m is the number of moles of gas and T is its temperature in kelvins. R is the gas constant, measured per mole. In Chapter 2, R was given as 8.314462 J mol^{-1}K^{-1}, based on the latest measurements. One of the many difficulties of the experiments made in the 19th century was of course the absence of a coordinated unit system to the extent that we have today (unless you still use the fps system of course), so the results of the different experiments had to be incorporated before arriving at a single (very simple) equation.

The remarkable thing about the gas law is that the volume of the gas becomes zero at $T = 0$, absolute zero. This is impossible for a real gas, so in practice all gases diverge from an ideal gas at sufficiently low temperature. This is because as the volume shrinks, the gas molecules come closer together and start to repel each other. Most gases in fact liquefy at low enough temperature, but the last to do so is Helium at 4 K[12]. The above equation, with $m = 1$, $P = 101325$ Pa and $T = 273.15$ K gives the volume of an ideal gas at standard temperature and pressure (0 °C, 1 atm). The result is 8.314462 x 273.15/101325 = 22.414 x 10^{-3} m^3, or 1 mole = 22.414 L. Real gases vary only slightly from the ideal volume an observation due to the Italian physicist Avogadro in 1811[15], with modern values shown in the table below.

Densities and molar volumes of some gases.

Gas	Formula	Molecular Weight	Density at 0°C kg m^{-3}	Molar volume m^3 kmol^{-1}
Acetylene	C_2H_2	26.040	1.170	22.26
Air		28.966	1.293	22.40
Ammonia	NH_3	17.031	0.769	22.15
Argon	Ar	39.948	1.784	22.40
Benzene	C_6H_6	78.110	3.486	22.41
Butane	C_4H_{10}	58.120	2.683*	21.66
Butylene	C_4H_8	56.110	2.504	22.41
Carbon dioxide	CO_2	44.010	1.977	22.26
Carbon monoxide	CO	28.011	1.250	22.41
Chlorine	Cl_2	70.906	3.213	22.07
Ethane	C_2H_6	30.070	1.357	22.17
Ethylene	C_2H_4	28.030	1.260	22.25
Helium	He	4.003	0.179	22.42
Hydrogen	H_2	2.016	0.090	22.42
Hydrochloric acid	HCl	36.461	1.630	22.37
Krypton	Kr	83.800	3.740	22.41
Methane	CH_4	16.044	0.717	22.38
Neon	Ne	20.179	0.900	22.42
Nitrous Oxide	N_2O	44.013	1.975*	22.28
Oxygen	O_2	31.999	1.429	22.39
Ozone	O_3	47.998	2.140	22.43
Propane	C_3H_8	44.097	2.020	21.83
Propene	C_3H_6	42.100	1.876	22.44

Measuring the World, by John Austin

Gas	Formula	Mol. Wt.	Density at 0°C	Molar Volume
Sulphur dioxide	SO_2	64.060	2.926	21.89
Toluene	C_7H_8	92.130	4.111	22.41
Water vapour	H_2O	18.016	0.804	22.41
Xenon	Xe	131.290	5.860	22.40

Most of the data are taken from[13] but for butane, the information did not seem correct and has been taken instead from[14]. The latter also suppied the density of Nitrous oxide, and has been corrected from 15 °C, using the gas equation.

Avogadro's work was not accepted by other scientists until 1858, 2 years after his death, although we now have irrefutable evidence. In the table, the molar volumes are all in the range 21.66 – 22.42, but these are measured values and may be subject to some experimental error. In particular, it is difficult to accept that butane is so low at 21.66 whereas the similar molecule butylene is close to the ideal gas value of 22.41. Likewise propane looks a little small at 21.83. Even in the case of butane, though, the value of 21.66 is only 3.4% below the ideal gas value. Note that most volumes are slightly lower than the ideal gas figure by about 1 or 2%, because the gas molecules tend to attract one another, reducing the volume. In the case of the noble or inert gases (helium, neon, argon, krypton, xenon), chemical interactions between molecules are likely extremely small, and the molar volumes are very close to the ideal value. For several measurements the volumes are slightly above ideal by 1 part in 3000, but this is likely to be experimental error.

The main advantage of working with gas volumes is that calculating the chemistry is easy. For example, burning 1 kmol H_2 requires ½ kmol of O_2 to form H_2O, so the ratio of *volumes* of $H_2:O_2$ is 2:1. Methane in natural gas is more complicated but essentially the same idea: $CH_4 + 2O_2 \rightarrow CO_2 + 2H_2O$. So 1 m³ of CH_4 reacts with 2 m³ of O_2 to form 1 m³ of CO_2 and 2 m³ of steam (if the temperature is high enough). In my scientific career, I have spent over 30 years doing research on the chemistry of atmospheric ozone. The ideal gas equation was probably the second most important equation in that time, after the governing equations of atmospheric dynamics.

13.4 The Number of Atoms per Mole

The number of atoms per mole is known as Avogadro's constant = 6.0221413×10^{23} (Chapter 2). The Avogadro constant, N_A, connects the atomic world with the macroscopic world, so is fundamentally important to a range of processes. It also means that N_A can be measured in a variety of quite distinct ways. In 1865 it was calculated from the distance between air molecules, and knowing the air density, Loschmidt found N_A[16]. Later, once the charge on the electron had been accurately determined, N_A was calculated from the total charge needed to electrolyse silver. Modern methods include using the mass of the electron and spectroscopic data. Currently, possibly the most reliable method is to use X-ray crystallography with silicon crystals. In principle, if the crystal mass is m and the density is d, the number of atoms in the crystal is $n = m/(dx^3)$ where x is the distance between the atoms. The number of atoms per mole, N_A, is n/M where M is the number of moles in the sample = m/W, where W is the atomic weight of the crystal. Hence $N_A = m/(dx^3) \times W/m = W/(dx^3)$. The final expression for the Avogadro constant N_A depends only on the atomic weight of the sample, its density and the distance between the atoms. In practice, it's slightly more complicated as the atoms are arranged in a particular pattern in the crystal structure. Nonetheless, a single measurement of the crystal structure spacing can be made to very high precision using X-Ray crystallography. It doesn't depend on the sample mass. Most recent high precision measurements of N_A have an uncertainty of about 30 parts per billion[17].

13.5 Other Molar Measures: Acidity and Catalysis

The usual measure of acidity is on the pH scale. Although slightly tangential to the rest of the chapter, the definition of pH is the H^+ ion concentration in kmol m^{-3}[18]. In the term pH, H is for hydrogen and the p signifies the logarithm to base 10. pH varies between 1 and 14, so the H^+ content varies between 10^{-1} and 10^{-14} kmol m^{-3}. Neutral is pH 7, pH 1 is a very strong acid, and pH 14 is a very strong alkali. Natural water is slightly ionised: the molecule breaks down to form H^+ and OH^- ions in equal, very small quantities (10^{-7} kmol m^{-3}). Vinegar is a weak acid (pH 4), blood is slightly alkaline (pH 7.4) and caustic soda (NaOH) is a strong alkali (pH 14). We detect an acid in our mouths by its sour taste, whereas an alkali tastes soapy. Chemically, our taste buds are detecting hydrogen and hydroxyl ions respectively. When an electrical charge is present, our taste buds interpret this as sourness, or acidity and chemically, we are detecting the presence of H ions.

It is of course entertaining to produce a battery, generating electrical power from fruits or vegetables[19]. To do this, we need a ready source of hydrogen ions, and anything with a sour taste can be used (Sainsbury's kiwi fruits are especially useful for this, as being abnormally sour). A small current of the order of a milliamp can be generated between two electrodes one made from zinc and the other copper. This is an often repeated experiment in school science classes, and shows the connection between pH, here, and electricity in Chapter 12.

Finally, for completeness we include the SI derived

unit of katal (abbreviation kat) which is the unit for catalytic activity and is equivalent to moles per second. It is a specialist unit used in biochemistry. Although it had been used for decades, it only became an SI unit in 1999[20].

References

[1] Antoine Lavoisier, Wikipedia, 1 April 2014.
http://en.wikipedia.org/wiki/Antoine_Lavoisier, accessed 4 April 2014.
[2] Carbon-12, Wikipedia, 24 March 2014,
http://en.wikipedia.org/wiki/Carbon-12, accessed 4 April 2014.
[3] Avogadro constant, Wikipedia, 2 April 2014,
http://en.wikipedia.org/wiki/Avogadro_constant, accessed 4 April 2014.
[4] Example: Periodic table of chemicals, TikZ, Texample.net, Griffin, I., 20 December 2009, http://www.texample.net/tikz/examples/periodic-table-of-chemical-elements/, accessed 4 April 2014.
[5] Periodic table, Wikipedia, 30 March 2014,
http://en.wikipedia.org/wiki/Periodic_table, accessed 4 April 2014.
[6] WebElements: the periodic table on the web, Webelements, 2012,
https://www.webelements.com/, accessed 4 April 2014.
[7] Table of nuclides, Wikipedia, 27 March 2014,
http://en.wikipedia.org/wiki/Table_of_nuclides, accessed 4 April 2014.
[8] Island of stability, Wikipedia, 31 March 2014,
http://en.wikipedia.org/wiki/Island_of_stability, accessed 4 April 2014.
[9] Mass spectrometry, Wikipedia, 1 April 2014,
http://en.wikipedia.org/wiki/Mass_spectrometry, accessed 4 April 2014.
[10] Gas chromatography-mass spectrometry, Wikipedia, 10 March 2014,
http://en.wikipedia.org/wiki/Gas_chromatography%E2%80%93mass_spectrometry, accessed 4 April 2014.
[11] Gas laws, Wikipedia, 3 April 2014,
http://en.wikipedia.org/wiki/Gas_laws, accessed 4 April 2014.
[12] Periodic table of elements sorted by boiling point, Barbalace, K., EnvironmentalChemistry.com, 22 February 2007,
http://environmentalchemistry.com/yogi/periodic/boilingpoint.html, accessed 4 April 2014.
[13] Gases – densities, Engineering Toolbox,
http://www.engineeringtoolbox.com/gas-density-d_158.html, accessed 4

April 2014.
[14] Gas encyclopedia, Air Liquide, 2013, http://encyclopedia.airliquide.com/encyclopedia.asp, accessed 4 April 2014.
[15] Avogadro's law, Wikipedia, 1 April 2014, http://en.wikipedia.org/wiki/Avogadro%27s_law, accessed 4 April 2014.
[16] Avogadro constant, Wikipedia, 2 April 2014, http://en.wikipedia.org/wiki/Avogadro_constant, 4 April 2014.
[17] Andreas, B. et al., 2011: An accurate determination of the Avogadro constant by counting the atoms in a 28Si crystal, Phys. Rev. Lett., 106, 030801, doi: 10.1103/PhysRevLett.106.030801, 18 January 2011, http://arxiv.org/abs/1010.2317, accessed 4 April 2014.
[18] PH, Wikipedia, 26 January 2014, http://en.wikipedia.org/wiki/PH, accessed 4 April 2014.
[19] Lemon battery, Wikipedia, 3 April 2014, http://en.wikipedia.org/wiki/Lemon_battery, accessed 4 April 2014.
[20] Katal, Wikipedia, 18 April 2014, http://en.wikipedia.org/wiki/Katal, accessed 24 April 2014.

Measuring the World, by John Austin

Measuring the World, by John Austin

14. Light Brightness

14.1 Introduction

Mankind has been strongly influenced by light generated by a wide range of phenomena. Clearly, we recognise different sources of light from a candle flame, the sun, an aurora, a firefly and a laser as well as the electric light bulb. The usefulness of the light depends on its colour (wavelength) as well as brightness. Over the centuries, explanations of how these different light forms arise have been developed and in some cases applied to practical use. For example, we now know that the bright light from the sun arises from nuclear fusion[1], which leads to high surface temperature of about 6000 K and re-radiation according to the laws of physics. In the case of the incandescent light bulb, the filament is raised to about 3000 K[2] and it then radiates energy (which we see partly as light) at the redder end of the spectrum. A firefly display is an example of bio-luminescence[3] – chemical reactions in the insects abdomen generate flashes of light. In tropical and sub-tropical countries where the insects are present, children collect enough of them in a glass jar, and together they provide a pale light. The aurora which appears in the skies of high latitudes is a similar level of brightness, but arises from electron transitions in atmospheric molecules at specific wavelengths. Lasers also produce single wavelength transitions, contrasting with the broad range of wavelengths from the sun, candles and incandescent light bulbs.

Measuring the World, by John Austin

Modern physics recognises the wave-particle duality of light. In other words, light behaves like waves, for example in reflection and refraction, described in § 14.2-14.3. In other circumstances, light behaves like particles. Whether light was a particle or wave exercised the minds of the greatest scientists such as Newton and others from the 17th century onwards. It was not until the early 20th century, after the development of quantum mechanics, that the true nature of electromagnetic radiation (of which light is a small part) became fully appreciated. In one of the classic experiments, Young's double split experiment, light is incident on two slits and a diffraction pattern is produced, described in § 14.4. A similar pattern is produced even if electrons are passed through the slits, yet conventionally they are clearly considered as particles. As a type of light relief (yuk!) the behaviour of moths near a light are discussed. Have you ever wondered why insects tend to spiral in towards a light? One possible reason is revealed in § 14.5.

Having thus established the main properties of light we move onto more practical matters in § 14.6 to define the units of luminous intensity. This will enable the reader to compare the many different type of light bulbs that now exist to allow the easy determination of the desired lighting brightness for a given sized space. For the more exacting conditions light meters can be used, as described in § 14.7, and the relationship to photography is discussed. The many different types of light bulbs are compared in § 14.8-14.9. This has become an area of rapid technological change where a simple grasp simple calculations can answer otherwise difficult lighting questions, as explained in § 14.10.

14.2 Reflection of a light beam

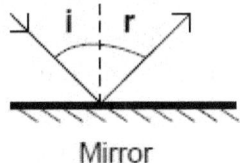

Mirror

Many surfaces such as glass or metal will reflect light. This is of course a well-known phenomenon which we see every day, literally! Modern mirrors are efficient in that the reflected light is almost as bright as the incident light. The angle of incidence, i, and the angle of reflection, r, are the same. In the diagram, these angles are measured from a line drawn perpendicular from the mirror surface. Measurements show that $i = r$[4]. Also, the image is reversed left to right. Curved mirrors, in which the incident beam is focussed to a point are useful in astronomy in telescopes to image distant objects[5].

14.3 Refraction of a Light Beam

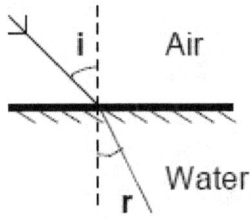

Refraction occurs when light transfers from one medium (such as air) to another (such as water or glass)[6]. Refraction is a bending of the direction of the light beam and commonly occurs between air and glass (producing for example a rainbow by a prism) or air and water (again, a rainbow by raindrops in the sky). The effect of refraction in water is to make objects appear at lower depths than they are in reality, a well-know effect in a swimming pool or stream. The relationship between the angles i and r is given by Snell's law $n_1 \sin i = n_2 \sin r$. n_1 is the refractive index of material 1 and n_2 is the refractive index of material 2. for example, for a vacuum, $n_1 = 1$ exactly. For other common substances, the refractive indices are: air , $1 + 2.93 \times 10^{-4}$; water, 1.333; glass, 1.46 to 1.62[7]. The highest refractive index for a uniform material is diamond at 2.42. However, it has been possible to construct metallic lattices, which have much higher refractive index, up to 38.6[8], for long wavelengths. The velocity of light in a medium is c/n where n is the refractive index and c is the velocity of light in a vacuum. n is always greater than or equal to 1, so light slows as it enters a more optically dense region.

14.4 Diffraction of a Light Beam

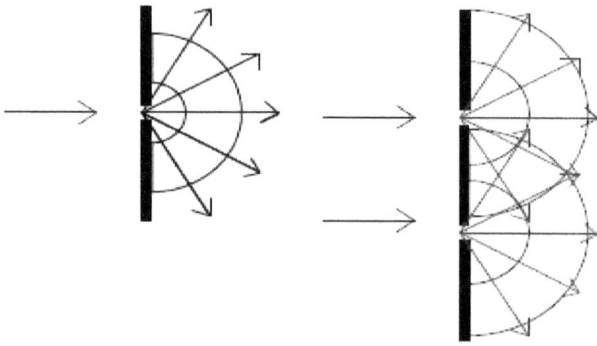

Measuring the World, by John Austin

If light is incident on a narrow slit (with size near that of that of the wavelength especially), light is 'diffracted' outwards[9]. The resulting diffraction pattern shows bright and dark regions radiating outwards with the details of the pattern depending on the wavelength of light. In the 18th and early 19th century, the properties of light were explored in considerable depth, and one of the classical experiments was Young's double slit experiment[10]. This produced an interference pattern in which the diffraction patterns from each slit overlapped. The above figure schematically illustrates the process of diffraction. On the left, light is incident on a single slit and radiates outwards from the slit, the diffraction pattern illustrated by the semicircles which indicate possible bright (or dark fringes). In the double slit experiment on the right in the figure, the diffraction patterns reinforce each other in a more complex mosaic of bright and dark fringes.

Although the above figure illustrates the passage of light through single and double slits, the same pattern also occurs for electrons, indicating that particles could also be considered as wave-like. If the incident rate of the electrons is reduced, in other words, the brightness of the electron source is reduced, only one electron could in principle be made to intercept the slit at a time. Yet the interference pattern remains. So no matter how few electrons pass through the system, it is not possible to determine which of the two slits an electron has passed through. This bizarre behaviour was not fully explained until the advent of quantum mechanics in the 1920s[11] in which the wave-particle duality of nature is built into the theory. In other words, all particles (even you!) have a wave-like equivalent. For a particle of very small momentum, such as an electron fired at a slit there is some uncertainty in its position (the so-called Heisenberg uncertainty principle) and we interpret the positional uncertainty as a wave. Diffraction patterns have a major practical application in determining the structure of crystals, and indeed were used to determine the structure of deoxyribonucleic acid (DNA) by Francis Crick, James Watson and Maurice Wilkins using X-rays[12]. The Nobel prize was awarded to Crick and Watson, and Wilkins, in 1962, and at that time the prize was not awarded to deceased scientists. Surely, Rosalind Franklin, who prepared the crystals and the diffraction images and completed some of the other major scientific work would have been a worthy co-winner of the prize, but she died prior to the Nobel award being made. Nonetheless, it appears that the scientific establishment, at that time in the UK, was distinctly chauvinistic[13], an embarrassment to current scientists no doubt.

14.5 Moths to a Flame

You will probably have observed moths flying around a lamp in ever decreasing circles. Why do they do it? Moths are perhaps confused about artificial light sources, which they could interpret as the sun or moon. The moth navigates by setting a fixed angle between the moon for example, and its flight direction. This works if the light source is a very long distance away. In the case of an artificial source, as it gets closer it needs to fly in a curve to maintain the angle constant as before. The result is that the moth spirals in towards the light until it collides with it. So much for one theory.

Apparently this has been investigated in detail, and the reality isn't quite so simple[14]. One simple suggestion is that evolutionary changes may have been triggered in moths to allow them to seek out mating partners. Once in a well-lit environment, they often find a relatively dark place to settle down, rather than continue the suicidal mission into the light.

Measuring the World, by John Austin

14.6 Luminous Intensity and brightness

The brightness of light bulbs is of major importance in the home and for industrial applications. The brightness of a light falls off quickly with the square of the distance, known as the inverse square law, but of course the total light emitted is still the same. The units of lighting[15] take account of these two aspects, the total light emitted in all directions, and the light brightness at a given distance from a light source. For example, to decide on the lighting for a given room, we need to decide the light brightness (light energy per unit area) at the sides of the room. Photography works on the principle of creating an image (electronic in the case of digital film, chemical in the case of traditional film), based on the total number of light particles (photons) from all distances incident on the film.

The *luminous intensity* of a light source is based on what used to be a standardised candle. In modern applications, the SI system defines the candela (cd):

The candela is the luminous intensity, in a given direction, of a source that emits monochromatic radiation of frequency 540×10^{12} hertz and that has a radiant intensity in that direction of 1/683 watt per steradian[16].

A steradian is what is known as a 'solid angle' in mathematics. All light sources present an area to the observer. The passage of light from this area to the observer is enclosed by a three dimensional cone, which is the solid angle. If the light source completely enveloped the observer, then the surface would be the surface of a sphere. To obtain the solid angle, the area involved is divided by the radius squared. In the case of the sphere of radius r surrounding the observer then the area would be $4\pi r^2$ and hence the solid angle would be $4\pi r^2/r^2 = 4\pi$ steradians. Although this sounds a little complicated, it is actually a straightforward extension of normal angles drawn on a 2-D surface. The 2-D equivalent of a sphere is the circle, and the total angle inside the circle is the length of the arc divided by the radius or $2\pi r/r = 2\pi$, and the angle is measured in radians. We also define a complete circle as 360°, so 2π radians = 360°, and in mathematics radians are often used instead of degrees. For small angles, radians have the useful property for example that *sin α ≈ α.* The frequency for the definition of the candela is equivalent to the colour green, near the middle of the visible spectrum.

In practice most lights are not monochromatic, but emit at a range of different wavelengths. The definition of the luminous intensity is then modified to correct for the human sensitivity to light at these different wavelengths. For the details see [15]. Lamps are usually rated by their sum over all solid angles. So a typical light bulb of, say, 100 cd would emit over almost a full 4π (≈ 12.6 steradians), giving about 1200 lumens (lm) of *luminous flux*, or in other words

$1\ cd = 1\ lm\ sterad^{-1}$.

As noted above, the apparent brightness of light, the intensity of light on a given surface, varies as the inverse square of its distance from the source, $b = E/r^2$, where E is the luminous flux and r is the distance. If E is measured in lumen (lm) and r is in metres, then b is in lux. If E is measured in lumen and r is in feet, then b is in foot-candles:

1 lux = 0.09290304 ft-candles exactly.

14.7 Measurements With Light Meters

Light brightness is measured in photography to determine the correct exposure. The use of such a meter is often thus expressed in exposure terms, e.g., *f8, 1/125 s*, which would be a typical daytime exposure for average sensitivity film (21 din, 100 ASA)[17]. With the rapid development of digital photography with automatic light metering, film speed or the electronic equivalent is becoming less recognised by most people. However, *f8, 1/125 s = 10* lux, approximately[18]. This is the brightness of the light averaged over the usual 24 x 36 mm image area of film.

14.8 Light bulbs

The electric light bulb[19] was invented by Edison in 1879, and, although the materials have undergone changes, it is remarkable that the design has remained almost unchanged until relatively recently. However, by modern standards, the incandescent light bulb is extremely inefficient converting to light only a small fraction of of the energy consumed. The vast majority of the rest goes wastefully to heat, which is why the bulb is too hot to be touched! For example, a typical 100 W bulb generates about 1160 lumens of light, for a total energy weighted towards the green of 1160/683 W or 1.7%. If the wavelengths are equally weighted, the energy efficiency is perhaps twice as much depending on the wavelengths included, but this still implies that the bulb generates 97% of its energy in heat, i.e. that amount outside the visible spectrum. Furthermore, their duration has been extended to only about 1000 hours at the voltage of the European power supply. The relatively short life of incandescent bulbs is due to the on-off cycle of heating and cooling which stresses the filament.

Lamps used for lighting streets are the familiar sodium discharge lamps which provide a characteristic yellow glow from the excitation of electron levels in sodium vapour. Low pressure lamps provide up to 200 lumens per watt[20], an efficiency of about 29%. High pressure lamps provide a brighter overall light, but require more power to operate and are rated at about 100 lumens per watt.

Other bulbs in common use are metal halide lamps[21]. These are gas discharge lamps and not to be confused with halogen lamps, which are incandescent bulbs. Although not as efficient as sodium lamps, they typically deliver 75 - 100 lumens per watt[21], More specialist lamps include mercury vapour lamps which provide output in the blue and ultra-violet for attracting insects for example. For these lamps, light production is up to 50 lumens per watt.

Compact fluorescent lamps[22] work on the same principle as a discharge lamp and are often suggested as substitutes for incandescent bulbs in the home. They have quite high efficiency, compared with incandescent bulbs, 50-70 lumens per watt – 9% efficient. They also have a long life: typically 10,000 hours for a 20 W, 1200 lumen bulb. However, they produce a bluish-white light and many people find this a little cold. Another problem with compact fluorescent bulbs is that they contain mercury, adding to the environmental problems on disposal.

With the range of energy efficient bulbs on the market, many governments have partial bans in place concerning the use of incandescent bulbs[23], as previously noted in § 12.13. Compact fluorescents are certainly a solution, but the future of lighting is likely to be using light emitting diodes (LEDs), as now discussed.

14.9 Light Emitting Diodes

With LEDs the light is emitted by solid state electronics [24], rather than the heating of a wire (incandescent lamps) or electronic discharge in a gas (sodium lamps, compact fluorescent). One of my first experiences of LEDs was in the mid 1970s in a pocket calculator which had only recently become available. The calculator was expensive, £30, which in today's money would be just over £200[25], yet you could buy such a calculator nowadays for about £5. It had 8 red LEDs which exhausted a set of batteries in just 1 hour of continuous use. Unfortunately, the only warning

Measuring the World, by John Austin

that the batteries were running down was that the calculator started making mistakes! So to test it periodically, I would put in a calculation for which I had memorised the result. If the calculator came up with the right answer, I could trust its subsequent calculations.

LEDs have come on a long way since then, and are starting to play a significant role in lighting. LEDs emitting in the infra-red and red had been available for some time, and other colours were produced by filters and fluorescence. This tended to lower the energetic efficiency. The breakthrough came with the invention of a bright blue LED by Shuji Nakamura[26]* and now all the colours can be represented by combining single wavelength light for red, green and blue part of the spectrum. Because no filters are required, a higher proportion of energy goes to light. Lamps are rated according to *efficacy* and *energy efficiency*. *Efficacy* is the light output in lm W^{-1}. With the lm definition based on 540 THz (green) light, the energy efficiency of LEDs at this wavelength is the efficacy divided by 683, which appears in the lumen definition. For other colours, the human response needs to be taken into account. So for example, as of 2012[24] the highest efficacy blue gave about 37 lm W^1, which combined with the human response of about 0.15[15] at this wavelength gives an actual energy efficiency (i.e. light energy as a fraction of the electrical energy consumed) of 37/(683 x 0.15) ≈ 0.35, or 35% energy efficient. Although higher performances are available for low powered LEDs, as of April 2014, typical efficacies for commercially available LEDs for the home are about 90 lm W^1 [27].

The full colour range mean that LEDs can be used for televisions[28] as well as lighting, and the lighting colour can be controlled precisely. Moreover, because LEDs are solid state devices they don't need time to warm up. They are therefore very controllable and can switch on and off instantly and also have a very long life time. For example, the owner of Terminal 5 Heathrow was embarrassed at the end of 2013, because they had no means of changing their light bulbs very easily[29]. The bulbs had remained in service for over 5 years without maintenance, despite almost continuous 24 hours per day use. One of the most public displays of LEDs is the NASDAQ display in Times Square, New York, which uses 19,000 LEDs[30]. The problem of achieving a bright blue LED[26] was solved using gallium nitride crystals with a special technique to reduce wafer imperfections which would otherwise absorb energy and heat up the wafer. LEDs which emit in the ultra-violet have also been invented. These can be used for example to purify small volumes of water for personal drinking.

"Light up the world" is on a charitable mission to replace lamps in the developing world with LEDs[31]. People in Nepal, e.g., typically use kerosene lamps for light. They're expensive on fuel, produce pollution in the home and provide limited light output. If the lamps are knocked over, the kerosene spills and people can get badly burnt. Furthermore, there is often a long distance to walk to buy the kerosene. In Sri Lanka, about 40% of burns are caused by the breakage of kerosene bottle lamps. LEDs have the promise of transforming our own lives, as well as the urban poor.

14.10 Calculating Comfortable Light Levels

Establishing proper light levels for any activity is important[32]. If lighting is too dim, eye strain can result and any activity or work that takes place might be inefficient. If lighting is too strong, headaches might result, especially under fluorescent lighting. Many of the health effects of lighting vary somewhat with the individual, but good estimates can be made as a starting point. Here I just give a quick example and the reader is referred to [32] for more details of comfortable light conditions. Let's examine a situation where, lighting for a kitchen is desired. The estimated required light levels are about 500 lux[32] to provide some judgements about textures and colours. Suppose the kitchen is lit by a single lamp in the centre of the room, and for argument sake the room is 3 m by 3 m. The room is chosen as square for convenience: these are approximate figures anyway, and precise details would require trial and error. The distance from the lamp could be up to 3/2 = 1.5 m, so the luminous flux of the lamp needed is 500 x 1.5^2 ≈ 1125 lm. Taking the performance of the lamp as 100 lm W^{-1}, we would need an 11 W LED lamp, or a 100 W incandescent. Now suppose the room were a bit larger, say 4 m, then the light needed would be 500 x 2^2 = 2000 lm. This is a very bright lamp and in practice we would be better off having 2 lamps of 1000 lm each, or some combination, although as they would be closer to the area of interest, a lower intensity, might suffice, say about two lamps of 750 lm. Overall, though, the main information you need is the size of the room, and the brightness of the lamps, measured in lumens. Similar calculations to the above are easy enough in the fps system, using the desired brightness in foot-candles, rather than lux, and measuring the size of the room in feet. However, in the process of doing this, units from different systems are being mixed. This is something of a howler for a schoolchild, so you shouldn't do it as an adult! More to the point, I now need my wife to agree to upgrade our kitchen light!

References

[1] Nuclear fusion, Wikipedia, 2 April 2014,
http://en.wikipedia.org/wiki/Nuclear_fusion, 4 April 2014.
[2] Color temperature, Wikipedia, 13 March 2014,
http://en.wikipedia.org/wiki/Color_temperature, accessed 4 April 2014.
[3] Bioluminescence, Wikipedia, 1 April 2014,
http://en.wikipedia.org/wiki/Bioluminescence, accessed 4 April 2014.
[4] Reflection (physics), Wikipedia, 19 March 2014,

http://en.wikipedia.org/wiki/Reflection_(physics), accessed 4 April 2014.
[5] Reflecting telescope, Wikipedia, 12 March 2014, http://en.wikipedia.org/wiki/Reflecting_telescope, accessed 4 April 2014.
[6] Refraction, Wikipedia, 3 April 2014, http://en.wikipedia.org/wiki/Refraction, accessed 4 April2014.
[7] Refractive index, Wikipedia, 3 April 2014, http://en.wikipedia.org/wiki/Refractive_index, accessed 4 April 2014.
[8] Metamaterial breaks refraction record, Physics World, Institute of Physics, 16 February 2011, http://physicsworld.com/cws/article/news/2011/feb/16/metamaterial-breaks-refraction-record, accessed 4 April 2014.
[9] Diffraction, Wikipedia, 31 March 2014, http://en.wikipedia.org/wiki/Diffraction, accessed 6 April 2014.
[10] Double-slit experiment, Wikipedia, 5 April 2014, http://en.wikipedia.org/wiki/Double-slit_experiment, accessed 6 April 2014.
[11] History of quantum mechanics, Wikipedia, 29 March 2014, http://en.wikipedia.org/wiki/History_of_quantum_mechanics, accessed 6 April 2014.
[12] DNA, Wikipedia, 5 April 2014, http://en.wikipedia.org/wiki/DNA, accessed 6 April 2014.
[13] Maddox, B., 2003: The double helix and the 'wronged heroine', Nature, 421, 407-408, 23 January 2003, doi:10.1038/nature01399, http://www.biomath.nyu.edu/index/course/hw_articles/nature4.pdf, accessed 6 April 2014.
[14] Why are moths attracted to bright lights?, Adams, C., The Straight Dope, 27 January 1989, http://www.straightdope.com/columns/read/1071/why-are-moths-attracted-to-bright-lights, accessed 6 April 2014.
[15] Candela, Wikipedia, 8 February 2014, http://en.wikipedia.org/wiki/Candela, accessed 6 April 2014.
[16] Units of luminous intensity (candela), BIPM, Sèvres, France, http://www.bipm.org/en/si/si_brochure/chapter2/2-1/candela.html, accessed 6 April 2014.
[17] Exposure (photography), Wikipedia, 13 March 2014, http://en.wikipedia.org/wiki/Exposure_(photography), accessed 6 April 2014.

[18] The ultimate exposure computer, Fred Parker Photography, http://www.fredparker.com/ultexp1.htm, accessed 6 April 2014.
[19] Incandescent light bulb, Wikipedia, 30 March 2014, http://en.wikipedia.org/wiki/Incandescent_light_bulb, accessed 6 April 2014.
[20] Sodium vapor lamp, Wikipedia, 31 March 2014, http://en.wikipedia.org/wiki/Sodium-vapor_lamp, accessed 6 April 2014.
[21] Metal halide lamps, Wikipedia, 11 March 2014, http://en.wikipedia.org/wiki/Metal-halide_lamp, accessed 6 April 2014.
[22] Compact fluorescent lamp, Wikipedia, 28 March 2014, http://en.wikipedia.org/wiki/Compact_fluorescent_lamp, accessed 6 April 2014.
[23] Bans of incandescent light bulbs, Wikipedia, 25 March 2014, http://en.wikipedia.org/wiki/Bans_of_incandescent_light_bulbs, 6 April 2014.
[24] Light emitting diode, Wikipedia, 6 April, 2014, http://en.wikipedia.org/wiki/Light-emitting_diode, accessed 7 April 2014.
[25] Historic inflation calculator: how the money has changed since 1900, R. Browning, This is money.co.uk, http://www.thisismoney.co.uk/money/bills/article-1633409/Historic-inflation-calculator-value-money-changed-1900.html, accessed 7 April 2014.
[26] "Brilliant!: Shuji Nakamura and the revolution in lighting technology", by Bob Johnstone, Prometheus Books, 2007, http://www.amazon.co.uk/Brilliant-Nakamura-Revolution-Lighting-Technology/dp/1591024625.
[27] New 9.5 Watt – B22 high output standard shape LED bulb, LED Hut, http://www.ledhut.co.uk/led-bulbs/b22-led-bulbs/10-watt-b22-high-output-standard-shape-led-bulb.html, accessed 7 April 2014.
[28] LED TV, Currys, http://www.currys.co.uk/gbuk/, accessed 7 April 2014.
[29] Heathrow seeks high-wire walkers to change light bulbs, Daily Telegraph, Travel News, 25 Nov 2013, http://www.telegraph.co.uk/travel/travelnews/10472481/Heathrow-seeks-high-wire-walkers-to-change-light-bulbs.html, accessed 7 April 2014.
[30] How the NASDAQ Times Square display works, M. Brain,

Measuring the World, by John Austin

HowStuffWorks.com, ttp://money.howstuffworks.com/nasdaq-marketsite-tower.htm, accessed 7 April 2014.
[31] Light up the world, official website, http://lutw.org/, accessed 7 April 2014.
[32] http://sustainabilityworkshop.autodesk.com/buildings/measuring-light-levels, accessed 7 April 2014.

* While this print version was being prepared, it was anounced that Nakamura and coworkers were awarded the 2014 Physics Nobel prize for their development of the bright blue LED!

Measuring the World, by John Austin

15. The Very Small and the Very Large

15.1 Introduction

There are several reasons to survey the very small and the very large in the same Chapter. The first is that there are often connections between astronomical measurements and subatomic physics. A second reason is that metric units are used in these regimes, even by the media that would otherwise use fps and even if the journalists misunderstand them. Because of the tendency for the media to quote more metric in these areas, I shall largely refrain from giving conversions to the fps system, except in a few cases. Exact conversions can be deduced from the material in the earlier chapters, but it is usually simpler just to accept new units. Continued use then creates familiarity which is one of the goals of this book. For example, for a large part of my adult life, I have thought of the radius of the Earth's orbit as 93 million miles, but I am now more comfortable with the number 150 million km[1], which has the added bonus of a quick calculation for the time taken for light to reach us from the sun = $150 \times 10^9/c$ = 500 seconds.

Another common aspect of the small and large is that measurements are typically far removed from human-sized units such as the metre and kilogramme, and multiple powers of 10 need to be used to describe them. Some of these are large multiples and are used only in a few disciplines. To write values concisely requires the use of quantities such as 10^{-21} or 10^{23}. Although the notation should be easily recognised by any high school student, there is still a tendency for the media to "dumb things down" by writing numbers in the form 1/1,000,000,... and 1,000,000,.... Personally I find it very tedious to have to count the zeroes – and mistakes can be made when there are many! Another pet irritation of mine with large numbers is the use of terms such as "septillion, octillion etc." for large numbers. Not only is it a pain to convert these rarely used terms into proper notation, it also concerns me that the author of the article I'm reading hasn't done the conversion properly either. I'm often left wondering exactly what the correct power of 10 is, and I'm then forced to check out the source material to be sure.

The chapter begins in § 15.2 with a section on very small objects, of the subatomic scale and in § 15.3, mass and energy units are introduced for these particles. Instrumentation for observing very small and atomic sized objects such as chemical structures is described in § 15.4 – 15.5. The connection to astronomical sizes, namely through cosmic rays is discussed in § 15.6, and the observing of the universe and astronomical properties are explored in § 15.7 – 15.10. The emphasis is on how easy and natural it is for for measurements to be expressed in the SI system, but with some additional units for convenience.

15.2 The Very Very Small

Atomic sizes are of order 100 pm (10^{-10} m), and in fact vary in diameter between about 62 pm (0.62×10^{-10} m) for helium and 520 pm for caesium[2]. However, almost the entire mass of an atom is concentrated within the nucleus. We don't have exact information about the size of the nucleus. It is often treated as a point particle, which would imply infinite density. Physics avoids infinities, so the idealised view would not be correct. Careful measurements have deduced that the proton radius is 0.8768 fm (0.8768×10^{-15} m)[3] and it follows that atoms will have nuclei

Measuring the World, by John Austin

ranging in diameter between 1.75 fm (hydrogen) and about 15 fm (uranium). Therefore, most of the atom, except for a very small nucleus should be empty space. For example, if we represented the radius of the atom as the distance from the sun (1.5×10^{11} m), the radius of the nucleus would be about 5000 km, broadly the radius of the Earth. In practice, the space between the nucleus and the outer edge of the atom contains the charges of the electrons spread out in all three dimensions. It is the charge which gives the impression that 'solid' objects are solid. Certainly, the idea of the atom as a nucleus surrounded by electrons like planets around a mini solar system is quite wrong! In the solar system, the space between planets is very nearly a vacuum, and the planets all orbit in approximately the same plane. The atom is three-dimensional and charge is distributed throughout the region between the nucleus and the edge of the atom.

It is not possible to observe the nucleus directly. This follows from the Heisenberg uncertainty principle[5], the foundation of quantum mechanics[6], which is our best theory governing the properties of the very small. The uncertainty principle can be written $\Delta x \Delta p \geq \hbar$. $\hbar$ is Planck's constant (Chapter 2) divided by 2π, Δx is the uncertainty in position and Δp is the uncertainty in momentum of the object being observed. Note that this does not express a fundamental problem with any instrument used to make the measurements: it expresses a fundamental limit with any measurement system. In particular, to find the state of a system (a 'quantum state') requires sampling the system with at least one photon, and the above inequality reflects the recoil momentum of a perturbed system. In other words, the act of observing interferes with any system and changes the system itself in the process.

The structure of the nucleus has been explored indirectly in particle accelerators, more recently in machines such as the Large Hadron Collider[7]. These studies confirm that the nucleus is indeed very small but not without structure[8].

Obtaining large scale evidence for the effects of the uncertainty principle is difficult. There have been some suggestions that *free will* is a consequence of the uncertainty principle[5], but this seems dubious. You could argue that the big bang itself is a quantum mechanical fluctuation (since zero energy is strictly forbidden by the uncertainty principle)[9]. I have seen at least one T-shirt with the logo 'the universe: the ultimate free lunch', and who can argue? Another often discussed topic is the famous thought experiment of 'Schrödinger's cat'[10]. This is a cat who's very life depends on a quantum mechanical fluctuation, and the thought experiment was designed to test the interpretation of quantum mechanics soon after the theory was formulated. This is entertaining but somewhat off my intended subject area, but check out the reference.

15.3 Mass and Energy of Sub-Atomic Particles

Sub-atomic particles are often unstable, and can form or disintegrate inside accelerators. Particles are often travelling close to the speed of light in physics experiments. To save the trouble of converting back and forth between energy and mass units it is often convenient to use energy units in place of mass even when the particles are at rest. The total energy is determined by the famous Einstein relation $E = mc^2$[11], which includes the energy at rest (i.e. due to the particles intrinsic mass) plus all other forms of energy including its kinetic energy. In the case where the the remainder of its energy is kinetic, the kinetic energy is the total less the rest mass energy. So the kinetic energy can be written $mc^2 - m_oc^2$. For a moving particle $m = m_o/(1 - v^2/c^2)^{½}$ and $mc^2 - m_oc^2$ reduces to the Newtonian kinetic energy $½m_ov^2$ when the velocity of the particle is much lass than c. A convenient energy unit for sub-atomic particles energy is the electron volt, eV, which is the energy acquired by an electron under the acceleration of 1 volt. 1 eV = 1.602×10^{-19} J, i.e. the charge on the electron times the accelerating voltage, 1 V. Accelerators use banks of batteries to accelerate charged particles. Many particles are single charged, so the total energy at the end of the acceleration is the same as the voltage imposed.

1 eV is a small unit of mass-energy, and MeV (10^6 eV) is a more practical-sized unit. For example, the rest mass of the proton is 938.27205 MeV[12] and that of the electron is 0.51099893[13] MeV = 9.1×10^{-31} kg approximately. The neutron mass is slightly higher than the proton at 939.56538 MeV. Measurement uncertainties are about 20 parts per billion. Approximate conversions from electronvolts to kilogrammes are 1 eV = 1.6x10-19 J = $1.6 \times 10^{-19}/c^2$ = $1.6/9 \times 10^{-35}$ = 1.78×10^{-36} kg.

As noted in §15.2, observations provide evidence that neutrons and protons (as well as a host of other fundamental particles) are each composed of 3 quarks[8, 15], which vary substantially in mass from the lightest (the 'up' quark, 2-3 MeV) to the heaviest (the 'top' quark, 173 GeV). Each quark has a charge in multiples of $e/3$, where e is the electron charge. However, quarks are only ever observed within particles (3 per neutron, proton or electron) and 2 per lepton. The leptons themselves[16] vary substantially in rest mass from the μ (207 m_e) to the τ (3477 m_e) where m_e is the electron mass. Considerable news was generated in July 2012 by the discovery of the 'Higgs boson', an important particle in the 'standard' theory of the atom, and whose presence lends mass to other particles[17].

At the opposite end of the mass scale are neutrinos[19], light particles produced from radioactive decay and other processes. The scientific evidence indicate that they possess a small but finite mass. Massless particles are possible, but they can only acquire an existence by travelling at the speed of light (e.g. photons) which is impossible for particles with any rest mass.

Measuring the World, by John Austin

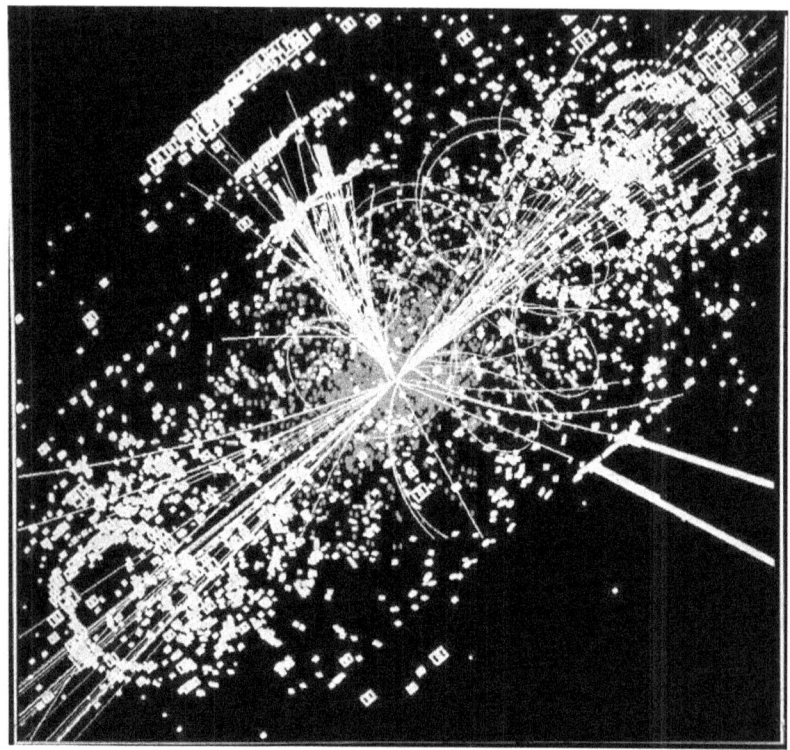

An example of simulated data on the Large Hadron Collider. Here, following a collision of two protons, a Higgs boson is produced which decays into two jets of hadrons and two electrons. The lines represent the possible paths of particles produced by the proton-proton collision in the detector with the specks indicating energy deposition. Image supplied by [18].

There are three neutrino types (electron-neutrino, v_e, tau-neutrino, v_τ, and muon-neutrino, v_μ) and experiments by for example the Planck collaboration have indicated that the sum of all three rest masses is less than 0.23 eV[20]. The lightest neutrino has rest mass of only about 0.05 eV. Although neutrinos are very light, their masses are very important because they could be sufficiently numerous to resolve some long-standing problems in astronomy, in particular why the expansion rate of the universe is increasing[21].

15.4 Microscopes for the Very Small

Although it is conventional to think of the power of a microscope as its maximum magnification, e.g. 300x, a proper measure is the 'resolving power'. This is the angle in radians (2π radians = $360°$) that just separates two points on the object being observed, or the minimum distance between two resolvable points. The limiting resolving power of a microscope is $1.22\lambda/D$ [22] where λ is the light wavelength and D is the objective diameter. In other words, microscopes are intrinsically limited by the wavelength of light itself. Comparing red light, $\lambda = 0.72 \times 10^{-6}$ m with blue light, $\lambda = 0.36 \times 10^{-6}$ m we see that the smallest resolvable angle is reduced by a factor 2 with blue light compared with red. For $d \approx 1$ mm, the effective magnification $\approx d/\lambda \approx 3000$ for blue light.

Further refinements in optics and microscope arrangements can increase the effective magnification. Continuing now in terms of resolving power, these improvements have led to a resolution of about 200 nm[22]. Improvements in resolving power can be achieved by exploiting the principles of quantum mechanics. Any particle with momentum p could be related to an equivalent wavelength by $\lambda = h/p$[23]. The De Broglie wavelength λ, is named after the French scientist who discovered the relationship. For example, in an electron microscope, for which the particles are electrons, λ could be as small as 0.1 nm. This leads to a further reduction in angle resolved by over 3600. The equivalent magnification is up to 3000 x 3600 or up to 10,000,000[24], and resolution as high as 50 pm. Unfortunately, not all samples are appropriate for examination. Whereas all samples reflect some light, and can be viewed with an optical microscope, providing a thin enough sample is obtained, the bombarding of the sample by high energy electrons can destroy it before a reasonable image has been obtained.

15.5 Microscopes for the Very, Very Small

For objects which can carry an electric current, the 'Scanning Probe' microscope can produce a resolution of a few picometres[25] or equivalent magnification of order 100 million, without destroying the sample. The microscope contains a probe which is brought to within 1 nm of the surface of the sample. A voltage is applied between the probe and the sample, creating a current in the sample surface, providing it conducts. This technique can be used for example to study the growth of silicon crystals to produce computer chips. The surface current is measured and kept constant while the probe moves over the surface. To maintain the current, the probe is raised or lowered, giving a monochromatic map of the sample surface.

For samples which don't conduct electricity, the 'Atomic Force' microscope can be used[26]. It has a probe which gently touches the sample surface. The mechanical fore on the probe is measured, and just like the scanning force microscope, the probe is adjusted to give a constant mechanical force. The position of the probe provides a monochromatic image of the sample surface and has a similar spatial resolution.

15.6 Cosmic Rays

Cosmic rays are streams of sub-atomic particles that bombard the Earth[27]. They arise from nuclear reactions in stars other than our own Sun, and can have extremely high energies. Cosmic rays are deflected by the stream of particles from the sun, known as the solar wind. The solar wind goes through an approximate 11-year cycle at times of high solar activity, cosmic rays become less intense. The last solar minimum was in about 2009, and cosmic rays were more intense than usual. The solar wind is nearing its peak or after its peak at the time of writing (April 2014) and cosmic rays are close to their minimum.

Cosmic rays can reach extremely high energies, about 3×10^{20} eV, far higher than in Earth-based particle accelerators. However, once the energy reaches 5×10^{19} eV, known as the Greisen Zatsepin Kuzmin (GZK) limit[28], particle kinetic energy is converted to mass according to the relation $m = E/c^2$. This is of course the usual Einstein mass-energy relation with mass, m, created from energy, E. Therefore above the GZK limit, particle fluxes should reduce substantially if not to zero. Nonetheless, the fact that some particles of this energy still occur may suggest weaknesses in our current physics theories. The GZK limit is several million times that achievable in the largest particle accelerators on Earth. Therefore cosmic rays continue to be investigated, amongst other things, to obtain access to very high energy particles, which cannot be artificially created. One of the outstanding challenges in the field is understanding the causes of energies close to and even beyond the GZK limit.

Cosmic ray particles are mostly electrons, protons and muons. Even 1 GeV particles can penetrate a 60 cm slab of steel. 98% of the particles are 'heavy particles' (protons 85%, helium nuclei 12%, muons and other heavy particles 1%). The remaining 2% of cosmic ray particles are electrons and positrons. In the atmosphere, most of the carbon is in the form of CO_2, with the carbon having an atomic weight of 12. Cosmic rays react with C-12 and change it to C-14, which is radioactive. This results in a small component of atmospheric carbon that can be used as a radioactive tracer or for dating carbon-containing specimens (Chapter 4). However, because the cosmic ray flux on the Earth varies with the solar cycle, corrections need to be made to the radiocarbon dates determined, to allow for the variable C-14 production rate.

The energy from cosmic rays is steadily absorbed by the atmosphere producing ions in a cascade process. Therefore, ion fluxes increase from lower to higher altitudes. For example, at 5000 m altitude, the intensity of ions is about five times higher than at sea level. In the upper atmosphere (20 – 25 km altitude) the exposure exceeds 2×10^5 particles $m^{-2}s^{-1}$, 1000 times the sea level value of only 200.

There has been endless, ridiculous debate about cosmic rays causes or contributing significantly to global warming. I shan't provide any references to this, as I don't want to give this issue any weight. It just makes me feel sad and uncomfortable that some people are prepared to latch onto anything to get away from the real issue of greenhouse gas concentrations in the atmosphere. It is easy enough to find studies on cosmic ray effects if you type "cosmic ray effects on climate" into an internet search engine. Suffice to say that a proposed mechanism via cloud formation is a non-starter in the atmosphere, as there are plenty of aerosols or particles to nucleate clouds, and ions are insignificant in the lower atmosphere for this. Of course there were periods when climate variations were correlated with solar variations, but that does not mean that there is a physical connection, and indeed the correlation has broken down in recent years.

What is known is that cosmic rays can damage electronic circuits[27], of particular importance to satellites, spacecraft and aircraft. One specific instance occurred in a Quantas flight in 2008 when a sudden electronics malfunction (most likely due to cosmic rays) caused the aircraft to drop hundreds of metres in altitude[29]. Aircraft systems have since been modified to prevent the effects of sudden power surges caused by cosmic rays. In 2010, the voyager 2 spacecraft malfunctioned due to a cosmic ray burst, but no serious damage occurred, and it continues to send back data.

Cosmic rays can cause adverse health effects, similar to that of radiation poisoning. Airline crews receive double the dose of ionising radiation received by the rest of the population, just through spending significant amounts of time at high altitude. Also, cosmic ray exposure is significantly higher for residents of high altitude cities such as Denver, USA and Mexico City, adding to their dose of UV from the sun. Cosmic ray exposure could be one of the most serious limitations to extended duration space flight[30]. In a return trip to Mars, astronauts would receive a radiation dose close to the limit of safety.

15.7 The Very Large

Astronomy is the domain of the vary large and very small. 'Astronomical' has entered our language in meaning something unimaginably large. Large, if not astronomical (!), sums of public money have been spent on telescopes and other tools, and scientists have responded in many ways, perhaps not least in supplying some of the most beautiful and fascinating photographs of the universe (e.g., from the Hubble Space Telescope[31]).

Most units in astronomy are closely related to metric units, with just a few relating to fps units. For example, for distances within the solar system, or for sizes of planetary bodies, miles are sometimes used. The conversion between miles and km (the preferred unit of course) is 1 mile = 1.609344 exactly (Chapter 5). Most of the time, however, metric units will be used here, and these are often quoted even by the US media. The distance from the Earth to the moon for example is now known quite accurately. The precision of the measurement was improved by placing a mirror on the moon during the Apollo programme, and the distance is determined by timing a laser beam[32]. This of course goes directly to the core of the definition of the metre. The current precision is about 2 cm (50 parts per trillion!) and as a result of these measurements, we now know that the moon is spiralling away from the Earth at the rate of 3.8 cm per year.

The next larger distance to be considered is the distance from the Earth to the sun[33]. This varies between 147,098,290 km at the minimum (perigee in the orbit, reached 2-5 January) and 152,098232 km at the maximum (apogee 4-7 July). In 2006, the BIPM defined the astronomical unit, ua, useful for solar system measurements, as the mean distance of the Earth from the sun, and is to be determined experimentally. The value is taken to be 149,597,870,691 ± 6 m[34], but in astronomical quarters is taken as a fixed distance of 149,597,870,700 m, i.e., 9 m further than the current mean measurement.

At larger distances still, the light year is a valuable astronomical unit, as the distance travelled by light in one year. Its value is the velocity of light x the length of a year in seconds = $2.99792 \times 10^8 \times 365.242 \times 86400 = 9.4605 \times 10^{15}$ m. Vast though this distance is, still larger units are required by astronomy. A practical unit is the parsec (abbreviation pc), dating back to the time when the distances of stars were first accurately measured. One parsec is the distance of an object which causes a parallax of 1 second of arc. Parallax is the change in the apparent angle of an object (in this case an when viewed on opposite sides of the Earth's orbit[35]. astronomical object) when viewed from a slightly different position. In the diagram, the Earth is shown

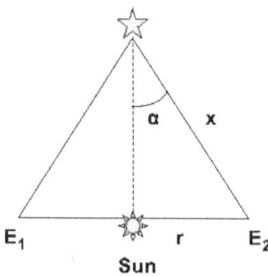

in two different positions E_1 and E_2 at opposite sides of the orbit around the sun. The angle of view in the two cases is slightly different and α is the parallax angle. By simple geometry *sin α = r/x* where *r* is 1 *ua* and *x* is the distance to the star. Since α is small, $\sin \alpha \approx \alpha$ (radians), so *x = r/α*. For α = 1 second of arc, α = 1/60 x 1/60 x π/180, and r = 1.49598x10^{11} m. The distance in light seconds is *r/(αc)* = 1.49598x10^{11} x 60 x 60 x 180/(π x 2.99792x10^8), and since we know that the Earth is about 500 s from the sun, *x* can be approximated by 500 x 60 x 60 x 180/π light seconds. In light years (ly) the distance is 1.49598x10^{11} x 60 x 60 x 180/(π x 9.4605x10^{15}) ≈ 3.2616 ly. While the parsec is a popular unit among astronomers, it is not that much bigger than a light year, so tends to add unnecessarily to scientific jargon, particularly now that the distances of relatively close objects are known to reasonable precision.

The light year can be small by astronomical standards, even though it would take a jet liner 1.3 million years to fly 1 ly at its cruising speed of 800 km h^{-1}. The Voyager spacecraft, launched in the late 1970s are still only 0.002 ly from Earth[35]. The nearest star is 4.3 ly away[36], and our galaxy (the milky way) is about 110,000 ly in diameter[37], bulging as much as 10,000 ly in the galactic centre. The solar system is about 27,000 ly from the centre of the galaxy, with an orbital period of about 240 Ma. The nearest galaxy (Andromeda) is 2,500,000 ly away[35], and the edge of the observable universe is about 46 x 10^9 ly across, very large indeed!

15.8 Observing the Universe

Many centuries ago, scientists used telescopes which were sensitive only to the optical region of the electromagnetic spectrum (Chapter 5). Nowadays, observations are made throughout the spectrum from gamma rays to radio wavelengths[38]. The table below shows the use to which each part of the EM spectrum is applied.

Band	Wavelength	Frequency (Hz)	Topics
Radio	> 1 mm	< 3×10^{11}	Pulsars, quasars, gas clouds orbiting the milky way
infra-red	720 nm – 1 mm	3×10^{11} – 4.2×10^{14}	Stars forming, relatively cool stars, planets
Visible	360 -720 nm	$4.2 – 8.4\times10^{14}$	Planets, stars, galaxies, asteroids, comets
UV	10 – 360 nm	8.4×10^{14} - 3×10^{16}	Hydrogen gas between stars; the sun
X-rays	0.1 – 10 nm	3×10^{16} - 3×10^{18}	The sun's corona, quasars, disks of matter around black holes
Gamma Rays	< 0.1 nm	> 3×10^{18}	Collapsed stars, matter-antimatter annihilations

Although other forms of telescope are common, optical telescopes still provide significant science. See e.g. [39], while as noted earlier, the Hubble Space Telescope has provided iconic images[31]. To make the most of observing conditions on Earth, telescopes need to be placed at high altitude to avoid the influence of pollution and interference from water vapour. It has become a technological challenge to build telescopes with a collecting mirror larger than about 8 m in diameter[40]. One way around this is to merge data from multiple telescopes using computer techniques and the combined signals are the equivalent of a larger telescope. So, for example, 4 telescopes of 5 m diameter would provide the same total light as a single 10 m telescope. Another approach is to use many mirror panels on each telescope and control each panel independently to minimise unwanted noise. An example of the former is Mt. Paranal in Chile, where there are 4 coordinated

Measuring the World, by John Austin

telescopes giving an effective diameter of 16 *m*. An example of the latter is in Hawaii where the Keck I and II telescopes are operating at an altitude of 3500 m and each contains 36 hexagonal-shaped mirrors.

Positions of astronomical objects are determined by measuring angles relative to the Earth's axis of rotation[41], and taking objects as residing on the celestial sphere. For example, the pole star is very close to the celestial north pole: objects in time-lapse photography are seen to rotate about it. The angle from the celestial equator is known as the declination, and is measured in degrees, minutes and seconds. It is analogous to latitude on the Earth's surface. The analogous measurement to longitude is known as right ascension and is measured Eastward from the spring equinox. So, the two angles together with the distance of a given object is sufficient to define its three-dimensional location in the universe.

15.9 The Doppler Effect and the Age of the Universe

As noted in § 10.10, the Doppler effect causes a change in the frequency of light or radiation emitted by an object. This can be used for astronomical objects to determine their velocity relative to us. To do this, the full spectrum of a source needs first to be observed. As first noted by Joseph von Fraunhofer[42], the spectra of all objects contain lines for where there is no radiation received by us. Essentially, the radiation has been absorbed by interstellar gases (especially hydrogen). These lines are now known as Fraunhofer lines in honour of their discoverer. Moreover, the lines have shifted in frequency by the time they have reached us. This implies that the source is moving relative to us. Since we can find the correct wavelengths when at rest in the laboratory, the velocity of the source relative to us can be determined. The Fraunhofer lines are consistently found to shift to the red (lower frequency), implying that most objects are moving away from us. The absolute distances of objects can be determined using "standard candles" in galaxies with high redshift[21]. These standard candles have have a known luminosity, because of their physics, and their brightness can be used to deduce distance by the inverse square law. It is found that the velocities of galaxies are proportional to the distance, with the constant of proportionality, the Hubble constant, H_o[43]. Unfortunately, recent measurements have shown some differences, but the current best-known value is 67.8 ± 0.8 [44] km s^{-1} Mpc^{-1}. Although common place, this is an obscure unit, partly because of the obscurity of the parsec. To convert to proper units, H_o = 67.8x10^3/(3.2616x10^6 x 9.4605x10^{15}) = 2.1973x10^{-18} s^{-1}. The age of the universe is then approximately $1/H_o$ = 4.551x10^{17} s = 14.42 billion years. Hence, the expansion rate of the universe would imply an approximate age of the universe, of just over 14 billion years. In practice, this figure is slightly too old, as the universe went through a rapid expansion in the early stages. A more considered figure arrives at the value of 13.80 ± 0.04 billion years[44,45], 3 times the age of the solar system (4.5 Ga).

15.10 The Sun and Other Stars

The diameter of the sun is 1,392,684 km[46], about 109 times the radius of the Earth. The mean rotation period is 25 days at its Equator and 28 days in its middle latitudes. Its stronger gravity and slower rotation rate produces only a small difference between equatorial and polar radii. The rotation rate is revealed in the properties of the heat and light emitted, as the sunspots move. Although sunspots appear dark (and hence emit less light than the surface overall), they are surrounded by bright regions (faculae), and the average over both sunspots and faculae is actually higher than the unperturbed surface. Consequently, solar output is higher

Measuring the World, by John Austin

when the sun is more active, that is, when there are more sunspots present. There is an approximate 11-year cycle in solar activity, but the largest fractional variation in activity occurs in the ultra-violet and shorter wavelengths, which constitute only a very small part of the total energy output from the sun. That total energy output, the Total solar Irradiance (TSI) varies only by about 0.1% during the cycle [47]. TSI is measured at the top of the atmosphere, and is the energy received from the sun when the sun is directly overhead. This means that the total output from the sun, the *solar luminosity*, which is radiated in all directions is $P = 4\pi A^2 \times TSI$ where A is 1 Astronomical Unit. For TSI = 1366 W m^{-2}, and $A = 1.49598 \times 10^{11}$ gives $P = 3.842 \times 10^{26}$ W. A slightly different value is often quoted, 3.839×10^{26}, i.e. 0.1% less, presumably based on a lower TSI.

Sunspots are connected with magnetic fields near the surface of the sun. The normal surface has a magnetic field of 50 – 400 µT[46], compared with 25 – 65 µT, for the Earth's magnetic field[48]. However, in the sunspots, the magnetic field is about 2500 times higher than on Earth[49], which would put the field at about 0.1 - 0.2 T. Solar flares appear periodically as bursts of light and matter from the surface. The bursts last from tens of seconds to tens of minutes. If directed towards Earth, the Van Allen Belts[50] deflect the majority of the charged particles, protecting our environment to a degree. Especially intense bursts, though, can disrupt communications and electronic equipment. The sun is an ordinary star in many respects, so there is every reason to suppose that similar activity occurs in most stars.

Using the Earth's orbital information, we can deduce other data about the sun. From the gravitational force on the Earth, we find $GMm/A^2 = mv^2/A$ where G is the gravitational constant, M is the mass of the sun, m is the mass of the Earth, and v its velocity in orbit. A is 1 ua. From this we find $M = v^2A/G$. Since $v = 2\pi A/y$ where y is one year in seconds, $M = 4\pi^2 A^3/(Gy^2)$. Of course this shows, incidentally, as M is of course the same for all the planets, that the period of the orbit of each planet squared is proportional to the cube of the distance from the sun, one of the laws that Kepler deduced from planetary observations[51], which allowed Newton to discover his law of gravitation. Putting all the values into the expression for M, $M = 4 \times 3.14159^2 \times (1.49598 \times 10^{11})^3 / (6.6738 \times 10^{-11} \times (365.242 \times 86400)^2)$ and using our trusty calculator, we get a values of $M = 1.9887 \times 10^{30}$ kg. This is quite a high precision value but is limited rather by the precision to which G is known: 1 part in 10,000, or 2 in the last figure. A similar calculation can be done using the moon orbiting the Earth, to find the mass of the Earth. Or as was done in § 9.7, the known value of the acceleration at the surface of the Earth can be used to calculate the Earth's mass. The net result is that the sun is 333,000 times more massive than the Earth. Combined with the sun's radius, we can calculate its mean density as 1406 kg m^{-3}. This is somewhat lower than the Earth's 5513 kg m^{-3}, calculated from the Earth's mass and radius in the same way. However, the sun is composed almost entirely of hydrogen and helium. At 0 °C and 1 atmosphere pressure, the density of hydrogen is 0.09 kg m^{-3} (§ 13.3), which indicates that the gases are under extremely high pressure due to gravitational compression.

    The luminosity of stars is of major importance to anyone observing the night skies. The brightness of stars is determined by their absolute brightness and their distance from Earth. Obviously, a close pale star could outshine a distant, bright one. These two perspectives are encompassed into two terms: the *apparent magnitude*, and the *absolute magnitude*[52], the brightness viewed from a standard distance of 10 parsecs (32.6 ly). to calculate the absolute magnitude, the brightness is assumed to fall off as the square of the distance. For example, if a star is of brightness b at 1 ly, then at 2 ly, its brightness will be $(½)^2 b = b/4$. In practice, the magnitude scale is reversed, and logarithmic. Bright stars have lower values and pale stars have higher values.

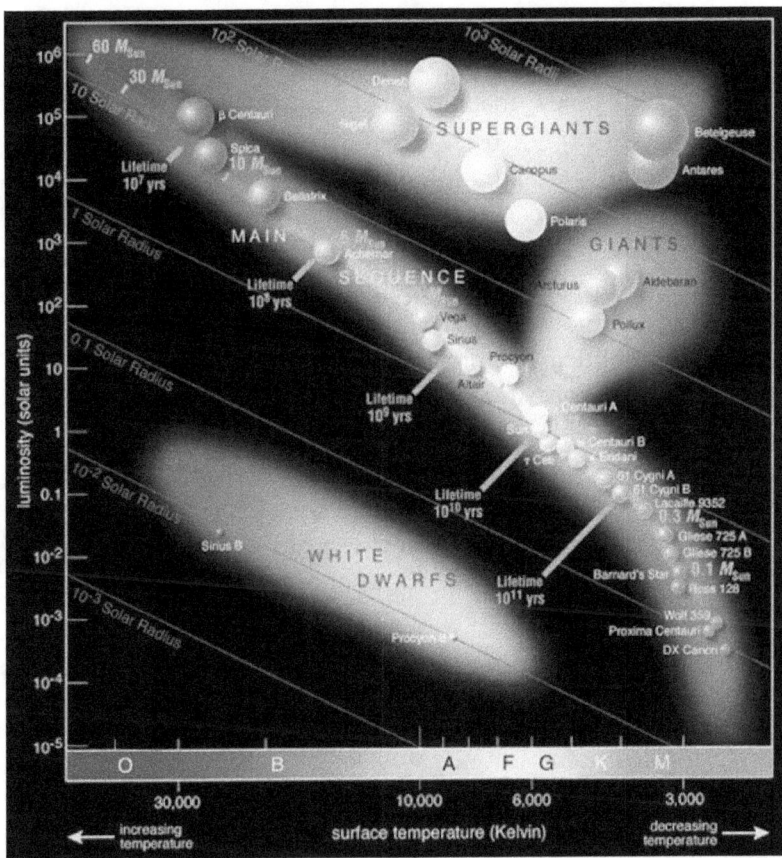

The Hertzsprung-Russell diagram classifies stars according to their surface temperature and luminosity, with most stars including the sun appearing on the "main sequence". Image supplied by the European Southern Observatory[54].

For our own star, the Sun, the luminosity is 3.842×10^{26} W as noted above. Apparent magnitude is -26.74[53]. Obviously it's very bright, as it is so close! Using the definition of star magnitude, we can calculate the apparent magnitude of the sun if it were at 10 parsecs,

$$M = M_o + 5\log_{10}(32.616cy/a_u)$$

where M is the new magnitude of the sun, M_o is its apparent magnitude (-26.74), $32.616cy$ is the distance of 10 parsecs where c is the speed of light and y the number of seconds in a year; a_u is the astronomical unit. Working through this, we find that $M = 4.83$, in other words, the sun's absolute magnitude is 4.83. We can also look at the information from another point of view, and calculate the luminosity for a star of given absolute magnitude. For a star of absolute magnitude M, this method calculates the luminosity as $3.28 \times 10^{28} \times 10^{-2M/5}$ W. Apparent magnitudes of up to 31.5 are detectable using the space telescope[53] and in principle, the faintest stars have absolute magnitude of 16. It is thought that a lower equivalent luminosity would not sustain the fusion reactions in the stellar core. Of the naked eye stars, those with highest luminosity include Deneb and Rigel, at absolute magnitudes -8.38 and -7.84[55]. The apparent brightest star, Sirius, has apparent magnitude -1.47 and absolute magnitude 1.4.

Star colours reveal their surface temperature, by the laws of 'black' body radiation. The peak in electromagnetic output, for example occurs at a wavelength given by the Wien displacement law[57], and this gives stars their characteristic colour. Surface temperatures lie between 2500 and 50,000 K, although in star cores where the fusion reactions take place, the temperatures are typically 10^7 K. Some of this information is contained in the Hertzsprung-Russell diagram shown in the above diagram[54]. The Sun is a star of typical mass, at 2×10^{30} kg. Other stars are for example α-Centauri 1.08 solar masses, Rigel 3.5 solar masses. Surface temperature and star mass depend on the evolutionary stage of the star[56]. Suffice to say that our orange-coloured sun (surface temperature 5700 K) is an average star of average temperature and mass. It has a likely future lifespan of several billion years, after which it will become a red giant and the

Measuring the World, by John Austin

Earth will likely be engulfed by an expanding sun.

References

[1] Astronomical unit, Wikipedia, 7 April 2014, http://en.wikipedia.org/wiki/Astronomical_unit, accessed 8 April 2014.
[2] Atom, Wikipedia, 24 March 2014, http://en.wikipedia.org/wiki/Atom, accessed 8 April 2014.
[3] The proton shrinks in size, G. Brumfiel, Nature News, 7 July 2010, http://www.nature.com/news/2010/100707/full/news.2010.337.html, accessed 8 April 2014.
[4] Atomic nucleus, Wikipedia, 21 February 2014, http://en.wikipedia.org/wiki/Atomic_nucleus, accessed 8 April 2014.
[5] Uncertainty principle, Wikipedia, 23 March 2014, http://en.wikipedia.org/wiki/Uncertainty_principle, accessed 8 April 2014.
[6] Quantum mechanics, Wikipedia, 6 April 2014, http://en.wikipedia.org/wiki/Quantum_mechanics, accessed 8 April 2014.
[7] Large Hadron Collider, Wikipedia, 8 April 2014, http://en.wikipedia.org/wiki/Large_Hadron_Collider, 8 April 2014.
[8] "The Quark and the Jaguar: Adventures in the simple and the complex", Murray Gellman, Published by W.H. Freeman & Co., New York, 1994.
[9] Is our universe the ultimate free lunch, M. Livio, A curious mind, 13 November 2012, https://blogs.stsci.edu/livio/2012/11/13/is-our-universe-the-ultimate-free-lunch/, accessed 8 April 2014.
[10] Schrödinger's cat, Wikipedia, 29 March 2014, http://en.wikipedia.org/wiki/Schr%C3%B6dinger's_cat, 8 April 2014.
[11] Mass-energy equivalence, Wikipedia, 27 March 2014, http://en.wikipedia.org/wiki/Mass%E2%80%93energy_equivalence, accessed 8 April 2014.
[12] Proton mass energy equivalent in MeV, NIST Reference on constants, units and uncertainty, http://physics.nist.gov/cgi-bin/cuu/Value?mpc2mev, accessed 8 April 2014.
[13] Electron mass energy equivalent in MeV, NIST Reference on constants, units and uncertainty, http://physics.nist.gov/cgi-bin/cuu/Value?mec2mev. Accessed 8 April 2014.

[14] Neutron mass energy equivalent in MeV, NIST Reference on constants, units and uncertainty, http://physics.nist.gov/cgi-bin/cuu/Value?mnc2mev. Accessed 8 April 2014.
[15] Quark, Wikipedia, 31 March 2014, http://en.wikipedia.org/wiki/Quark, accessed 8 April 2014.
[16] Lepton, Wikipedia, 23 March 2014, http://en.wikipedia.org/wiki/Lepton, accessed 8 April 2014.
[17] Higgs boson, Wikipedia, 6 April 2014, http://en.wikipedia.org/wiki/Higgs_boson, accessed 8 April 2014.
[18] CMS: Simulated Higgs to two jets and two electrons, L. Taylor, CERN (Conseille Europeen pour la Recherche Nucleaire), October 1997, http://cds.cern.ch/record/628469, accessed 8 April 2014.
[19] Neutrino, Wikipedia, 2 April 2014, http://en.wikipedia.org/wiki/Neutrino, accessed 8 April 2014.
[20] Planck collaboration, P.A.R. Ade et al., 2013: Planck 2013 results. XVI. Cosmological parameters, http://arxiv.org/abs/1303.5076, accessed 8 April 2014.
[21] The 4% Universe: Dark matter, dark energy and the race to discover the rest of reality, Richard Panek, Oneworld Publications, 2011.
[22] Angular resolution, Wikipedia, 20 March 2014, http://en.wikipedia.org/wiki/Angular_resolution, accessed 8 April 2014.
[23] Matter wave, Wikipedia, 5 March 2014, http://en.wikipedia.org/wiki/Matter_wave, accessed 8 April 2014.
[24] Electron microscope, Wikipedia, 26 March 2014, http://en.wikipedia.org/wiki/Electron_microscope, accessed 8 April 2014.
[25] Scanning probe microscopy, Wikipedia, 24 February 2014, http://en.wikipedia.org/wiki/Scanning_probe_microscopy, accessed 8 April 2014.
[26] Atomic force microscopy, Wikipedia, 28 March 2014, http://en.wikipedia.org/wiki/Atomic_force_microscopy, accessed 8 April 2014.
[27] Cosmic ray, Wikipedia, 4 April 2014, http://en.wikipedia.org/wiki/Cosmic_ray, accessed 8 April 2014.
[28] Greisen-Zatsepin-Kuzmin limit, Wikipedia, 29 March 2014, http://en.wikipedia.org/wiki/Greisen%E2%80%93Zatsepin%E2%80%93Kuzmin_limit, accessed 8 April 2014.
[29] Quantas flight 72, Wikipedia, 25 February 2014,

http://en.wikipedia.org/wiki/Qantas_Flight_72, accessed 10 April 2014.
[30] Health threat from cosmic rays, Wikipedia, 2 April 2014, http://en.wikipedia.org/wiki/Health_threat_from_cosmic_rays, accessed 10 April 2014.
[31] Hubblesite, Gallery, NASA, http://hubblesite.org/gallery/, accessed 10 April 2014.
[32] Apollo 11 experiment still going strong after 35 years, NASA JPL, news, http://www.jpl.nasa.gov/news/news.php?feature=605, accessed 10 April 2014.
[33] Earth's orbit, Wikipedia, 3 April 2014, http://en.wikipedia.org/wiki/Earth's_orbit, accessed 10 April 2014.
[34] Table 7. Non-SI units whose values in SI units must be obtained experimentally, BIPM, Sèvres, France, http://www.bipm.org/en/si/si_brochure/chapter4/table7.html, 10 April 2014.
[35] Parsec, Wikipedia, 27 March 2014, http://en.wikipedia.org/wiki/Parsec, accessed 10 April 2014.
[36] Astronomical distance scales, M. Guidry, Astronomy 162: stars, Galaxies, and Cosmology, University of Tennessee, Knoxville, http://csep10.phys.utk.edu/guidry/violence/distances.html, accessed 10 April 2014.
[37] Milky way, Wikipedia, 8 April 2014, http://en.wikipedia.org/wiki/Milky_Way, 10 April 20014.
[38] It takes more than one type of telescope to see the light, L. Mullen, NASA Space Science News, 6 April 2011, http://science1.nasa.gov/science-news/science-at-nasa/1999/features/ast20apr99_1/, accessed 10 April 2014.
[39] "Seeing and Believing: how the telescope opened our eyes and minds", Richard Panek, Penguin 1999.
[40] Engineering challenge: building a very large telescope, University of Washington Astrophysics Newsletter, http://www.astro.washington.edu/groups/observatory/newsletter/TJO_Fall08Winter09_george.pdf, accessed 10 April 2014.
[41] Right ascension, Wikipedia, 21 January 2014, http://en.wikipedia.org/wiki/Right_ascension, accessed 10 April 2014.
[42] Fraunhofer lines, Wikipedia, 9 April 2014, http://en.wikipedia.org/wiki/Fraunhofer_lines, accessed 10 April 2014.

[43] Hubble's law, Wikipedia, 10 April 2014, http://en.wikipedia.org/wiki/Hubble's_law, accessed 10 April 2014.
[44] Planck collaboration: Ade, P.A.R. Ade et al., 2013: Planck 2013 results. I. Overview of products and scientific results, http://arxiv.org/abs/1303.5062, accessed 10 April 2014.
[45] Age of the universe, Wikipedia, 4 April 2014, http://en.wikipedia.org/wiki/Age_of_the_universe, accessed 10 April 2014.
[46] Sun, Wikipedia, 26 March 2014, http://en.wikipedia.org/wiki/Sun, accessed 11 April 2014.
[47] Solar variation, Wikipedia, 9 April 2014, http://en.wikipedia.org/wiki/Solar_variation, accessed 11 April 2011.
[48] Earth's magnetic field, Wikipedia, 17 March 2014, http://en.wikipedia.org/wiki/Earth's_magnetic_field, accessed 11 April 2014.
[49] The sun and sunspots, NOAA National weather Service Forecasts, Sioux Falls, SD, 1 February 2013, http://www.crh.noaa.gov/fsd/?n=sunspots, accessed 11 April 2014.
[50] Van Allen radiation belt, Wikipedia, 7 April 2014, http://en.wikipedia.org/wiki/Van_Allen_radiation_belt, accessed 11 April 2014.
[51] Kepler's laws of planetary motion, Wikipedia, 7 April 2014, http://en.wikipedia.org/wiki/Kepler's_laws_of_planetary_motion, accessed 11 April 2014.
[52] Magnitude (astronomy), Wikipedia, 25 March 2014, http://en.wikipedia.org/wiki/Magnitude_(astronomy), accessed 11 April 2014.
[53] Apparent magnitude, Wikipedia, 12 March 2014, http://en.wikipedia.org/wiki/Apparent_magnitude, accessed 11 April 2014.
[54] The Hertzsprung-Russell diagram, European southern Observatory, 19 June 2007, http://commons.wikimedia.org/wiki/File:ESO_-_Hertzsprung-Russell_Diagram_(by).jpg, accessed 11 April 2014.
[55] List of most luminous stars, Wikipedia, 22 March 2014, http://en.wikipedia.org/wiki/List_of_most_luminous_stars, accessed 11 April 2014.
[56] Hertzsprung-Russell diagram, Wikipedia, 27 March 2014,

Measuring the World, by John Austin

http://en.wikipedia.org/wiki/Hertzsprung%E2%80%93Russell_diagram, accessed 11 April 2014.
[57] Wien's displacement law, Wikipedia, 7 April 2014, http://en.wikipedia.org/wiki/Wien's_displacement_law, accessed 11 April 2014.

Measuring the World, by John Austin

Measuring the World, by John Austin

16. Unit Agreement

16.1 Introduction

I believe that a valid goal for global society is to settle on a common set of units for both technical, cultural and every day purposes. While many nations subscribe to this view without question, a few pay lip-service to it, by for example agreeing to go metric like the UK and then taking 50 years or more to do it[1,2]. Any reasons for extending the conversion to metric such as the cost associated with rapid change has now long since disappeared. Just like English has become the international language that everyone wants to speak, so the metric system has become the international standard for measurement. The fps system no longer has any dependence of its own but is defined entirely by the metric system, e.g., 1 inch = 2.54 cm (exactly), which is a bit like a spoken language having its own vocabulary but using the grammar of, say, English. I believe that it is perfectly plausible for the public to be fluent in both sets of units, as I have done, but in the long term I don't see the need. Having said that, bushels and the differences between dry and liquid measure do baffle me at times. If you're cooking for example, is the measure for flour a liquid or a dry volume measure? The answer is not as straightforward as it may seem. Those Americans I have met generally don't understand these subtleties anyway, so why do they continue to *prefer* these funny units? US engineers, as we have seen on several occasions, seem to live in a world of their own, as far as units are concerned.

Measuring the World, by John Austin

So, as long as you know the metric system inside out, there is simply no need for the fps system at all. As we have seen in this book, everyday calculations and beyond are consistently simpler and easier in the metric system. It has been designed that way after all! Yet the fps system is based on a conglomeration of poorly understood ideas collected over 2000 years. When a unit is unavailable, it borrows from metric anyway! What a shambles!

Changing over to metric is discussed in § 16.2 more from the British perspective perhaps. Getting confused over units is perhaps a bit of fun, and usually not much harm is done, but it's not always like that and § 16.3 points out some of the pitfalls. One of the common explanations for not fully converting to metric in Britain, is the cost involved. However, this is a disingenuous argument, as many countries have done so without major expense and in § 16.4 some rough cost estimates are given. Another reason for paralysis on units is the existence of advocacy groups[e.g.,3] which favour the use of antiquated units, yet apparently they did not put up much opposition regarding the UK currency, which was compulsorily decimalised in 1971. In § 16.5 we see how on different occasions mixing up units lost an expensive satellite, and led to an aeroplane running out of fuel in mid-flight. These losses are presumably considered a price worth paying for maintaining the illusion of democracy. In § 16.6 we see how in general terms other nations have coped with changing over to metric units, especially British Commonwealth countries, most of which have successfully adapted.

Finally, I don't think anyone would seriously ban poetic language if it involved the unit system! You would have to have a poor sense of humour to do so. However bad they are for physics and engineering purposes, fps units have provided a richness in our cultural heritage that no sensible person would like to see expunged. Phrases like 'miss by a mile', 'miss by an inch', 'a ton of bricks' etc., etc. would continue to be used. But this hasn't stopped ridiculous debate about the metrication of idioms[4]. I suppose it's easier worrying about trivia than solving the real problem, that British society is not consistent in its use of units - fps or metric. I do suggest, though, that if you do want to know quantitatively how much you've missed by to get you're metric tape measure out rather than rely on a standard idiom!

16.2 Changing from FPS to Metric Units

Why is it useful to change to metric? Because there is serious risk of miss-communication, there are extra costs to society in maintaining two sets of units, and because many modern visitors are baffled by fps units[2]. To foreigners, the UK must seem a bit backward-looking and that surely affects international trade to the detriment of the nation. In many ways, in the UK we have the worst of both worlds. We educate children in the metric system and then throw them into the real world where they have to cope in a mixed unit situation without having learned the fps system properly. Unfortunately, the information they pick up about the fps system in society is not coordinated like it would be in a science class. In time, our work force have a coherent appreciation of neither the fps system nor the metric system. I was educated entirely in the metric system and spent almost 10 years enduring the fps system in the USA. Since leaving the USA to return to the UK in 2013, i.e. for the last 12 months or more, except for members of the UK Metric association, I have not met a single British born person who unprompted by units would tell me their weight in kilogrammes or their height in metres. *Why?* The answer is usually in stones (weight) or feet and inches (height), and even people in the USA won't understand the former. Unfortunately, our health service seem still to use these funny units. In a recent (March 2014) television news item about rising child obesity levels, children's desired weights came out as a range in stones and pounds, even though the BMI on which the study is based uses

Measuring the World, by John Austin

child height in metres and weight in kilogrammes. I was mystified, and surely it would have been a lot clearer in kilogrammes. The Met. Office itself is entirely metric, at least I thought it was when I was working for it. Yet in a television forecast I have heard forecasters talk about rainfall amounts in inches, if a few cm are expected, and they seem to be unable to resist the call to Fahrenheit once the weather starts warming up. Wave height charts are consistently given in feet. Whatever is wrong with metres? Use half metres if it doesn't give enough precision. Interestingly, on the Met. Office Marine Forecasts Glossary[5], atmospheric pressures and wave heights are given in proper metric units (hPa and m), but wind speeds are given in knots, and visibility is in metres for fog, but nautical miles for mist!

The experiences of countries such as Australia and the UK are relevant regarding the change over of units. Britain once had a peculiar currency given to us by the Romans (12 pence per shilling and 20 shillings per pound). Who knows what the Romans were thinking of? They couldn't have been very smart at mathematics or they would have developed a different system than the one based on I, V, X, D, C etc. Our currency is now decimal (100 p per pound). The change occurred in 1971 and there was a lot of resistance at the time. I vaguely remember that at the time there was a little bit of confusion and fuss, but on the whole things went smoothly and within a few years everyone seemed comfortable with the new system[6]. Many European nations went through a similar experience in 1999[7] when they went over to using the Euro. Presumably most countries now prefer it, and the EEC could also make a significant positive contribution by helping with the unit fiasco in the UK. It could remove at a stroke one of the major excuses for not completing the metrication process -- the cost. As it is, Britain originally had a target for metrication by 1975[8], but it has been a half-hearted effort at best (§ 16.6), and many industries are permitted to continue in antiquated units. Australia's experience can certainly be learnt from as it went

Measuring the World, by John Austin

metric in 1988[9], having started after us. Again its progress is discussed in § 16.6. Closer to home, Ireland changed all its roads in 2005[10]. By contrast, while all UK roads are marked in mph and miles, the public is forced to operate with an inefficient dual unit system. In addition, most developing countries have made the decision of going metric from the very beginning.

At least the metric system is generally known here in the UK, unlike in the USA, where every measurement I made in my 10 years' residence had to be converted to fps for the benefit of the locals. I think most Americans probably wouldn't know how many centimetres are in a metre, but then not many people there would know how many quarts are in a bushel either. In principle, there is no harm in the US strategy. It can feed its own citizens a diet of rotten units, and to all intents and purposes ban the metric system entirely with my blessing. Difficulties arise, however, when its people travel to foreign climes and they can't make themselves understood, or understand the local language (words or units). Also, international trade is in international units and if the US is to sell its products to foreign countries, other than the most compliant ones that it can bend to its will, it needs to package its products in international units. Many industries there continue successfully and efficiently with antiquated units, particularly for those that remain entirely within national boundaries. However with a `flattening of the world'[11], industries are becoming increasingly global, requiring globally recognised units. The oil industry is one such global industry but it continues to quote oil volumes in barrels. You could argue that since everyone uses the same unit, it doesn't matter. However, few people I talk to have anything but the faintest idea of how large a barrel is (159 *l*). Of course those people are not in the oil industry, but not only is the world flat it is also connected. If I'm trying to estimate climate change, for example, I need to know how much carbon is being burnt and it takes a lot more effort to go from barrels of oil, rather than say tonnes or litres, for me to work it out.

Measuring the World, by John Austin

One global industry that would be affected by a change in units are the financial markets which are nearly always expressed in peculiar units reflecting the idiosyncrasies of the individual industry, e.g., US Dollars per Troy ounce for gold, but dollars per tonne for the price of copper. Ultimately, I suppose there is no harm in this – the financial dealers work in numbers rather than objects and concern themselves only with whether the numbers are increasing or decreasing. The issue is more that if the underlying industries were to comply more with the metric system, the trading figures would change. So, e.g., whereas gold is about $1200 per tr oz, the new figure might become something like $40 per gramme. Everyone involved would quickly get used to the new base numbers. It would add a uniformity to the market but the markets are somewhat passive users of the underlying data. There is no particular need to connect the price of gold with that of copper, except perhaps in `pairs trading' and such like. The market, though is sufficiently flexible to take the changes into its stride.

16.3 The Problems of Mixed Units

The presence of 'mixed' units is commonplace when fps units are used, like the USA, or in countries like the UK which have failed to complete the metrication process. Mixed units occur when you use ostensibly one system but borrow from the other as the mood (it seems) dictates. For example, the fps system confuses measurements of heat energy in one set of units (e.g. therms) and mechanical energy in another (foot pounds). Electrical energy doesn't exist in fps, so it uses the kW h, a variant from the SI system! US engineers often use a fundamentally different version of the fps system than the version that is more common in Britain, in which the pound is not a mass but a force called the pound force. US engineers then define slugs as masses. I'm frequently confused with fps in working out whether my numbers should be multiplied by g, the acceleration due to gravity, because the fps system has not been put together in a consistent fashion. In the SI system there is no such confusion – 1 kg is a mass and 1 N is a force. Another problem with the use of mixed units across related industries is a general failure to communicate. As noted in Chapter 12, for a long time in the UK, gas heating cost was conveyed in pounds sterling per therm and electricity in pence per kW h. You had to do a lot of arithmetic to conclude that heating a home with gas was a more efficient proposition than heating with electricity.

Mixed units are often just a nuisance: it means the employment and cost of extra resources to maintain the status quo. Why a country such as the UK can feel it can just waste these resources is not clear to me. The USA is wealthy enough to be able to waste its resources, perhaps. Sometimes, though, it can be dangerous such as an aircraft running out of fuel in mid-flight. Ironically, this occurred to a Canadian aircraft soon after the country went metric (§ 16.6). It would be absurd, though to conclude that changing units is inherently dangerous, but perhaps the country has still not entirely learnt the lesson from this. There is a short term cost of unit conversion, as discussed in § 16.4, but the long term benefits should be far greater. Further examples of serious costs are given in § 16.5.

16.4 The Cost of Becoming Completely Metric

One of the reasons for not going completely metric is the cost to our infrastructure, and the road signs are put forward as a prime example. However, careful examination shows that this reason is spurious. We can use the example of Ireland, which recently changed all its road signs to metric. Although distance markers were already in km for some years, the speed limits were changed during a weekend in late January 2005[12]. Taking account of the difference in road length in our two countries, the extrapolated cost for the UK would be about £20 million[13]. The UK Metric Association have estimated that the cost for all the distance signs would be about £60 million, by replacing signs over an extended period taking advantage of the normal replacement period. Although this might lead to some confusion, some compromise between immediate change, and gradual change might be more attractive than an immediate change for distances. Therefore the total cost is likely to be in the region of £80 million, although this figure is probably about 10 years out of date. I tried to put this in the context of actual road expenses and taxes received from motorists, but up to date figures appear to be a government secret. Suffice to say that in 2008, the total taxes levied on motorists was about £48 billion, about 0.2% of the estimated *total* cost of metricating the roads.

Government departments offer their own figures and arguments for the reason for non-conversion to metric. It appears to this writer that they are biassing the evidence they supply to avoid change. The Department for Transport (DfT) estimates the cost at almost 10 times the above figure, despite the Irish experience. In the one estimate I could find[15] figures seem to be plucked from the air, with even £100 million allocated for 'preparatory work'. All one needs to do is convert the speed limits, e.g. 20 mph – 30 km/h, 40 mph – 60 km/h etc., but the DfT claims this is not simple. Let's suppose, however, that they are not simpletons at the DfT, then it means instead that the DfT is trying to turn metrication into an exercise of reviewing all speed limits. There is no harm in reviewing speed limits. If it's so important, why isn't it being done anyway in the fps system of units? Indeed in the process of metrication speed limit review might be a worthwhile *benefit,* but let's not confuse necessary with desirable spending. To claim that speed limit review is necessary is disingenuous of the DfT and I believe is included to inflate the cost. Further, to claim £100 million for this is quite absurd. To make the pill even more bitter, the DfT also comment on the costs of metrication to business[15] which are all highly questionable, and in any case is not within the remit of the DfT. An earlier cost estimate from 1978, when metrication was considered more positively in the UK, finally being shelved through neglect, the DfT had a figure of about £8 million[8] for the conversion of the roads. Allowing for typical inflation of about a factor of 6 since then[16], you would obtain about £50 million, and then add a bit more for the additional work that is highlighted, you perhaps get to £60 million, not far from realistic estimate of £80 million from the UK Metric Association. According to the DfT the average cost now works out at about £1500 per sign. Is anyone in the government paying attention, or are they deliberately turning a blind eye? In any case, there is apparently already a contingency of almost £750 million in the DfT budget for road sign conversion[8]. The fact that it hasn't been spent just illustrates the insincerity of those charged with preparing the appropriate reports.

Other costs include glasses in public bars and milk deliveries. It can't cost much to put a half-litre line on a pint beer mug can it? Over the decades, doorstep milk deliveries have declined, and the number of glass milk bottles delivered has dropped to 2 million per day, compared with 40 million in the early 1990s[17], a drop of a factor of 20! Many people buy their milk in plastic disposable containers in the supermarket. A pint bottle must cost a few pence to make, and presumably there are few factories left that actually process the pint bottles. The cost of converting to litres must be under £1 million. So there is now no excuse for continuing to measure milk in pints and other drinks in litres. The supermarkets have coped with other drinks in litre bottles or containers without the need for subsidy, so they can do it with milk.

According to the above estimates, I would think that about £100-150 million would cover all the costs associated with completing the metrication process in the UK, with a significant amount of that going to the roads. However, there appears to be little rational debate, even within government departments, who seem to have exaggerated the costs significantly. I would like to see an open debate in the UK as to what the costs would actually be for us, and to what extent subsidies from the EEC might be available. Whatever the costs, I believe also that this cost be born by general taxation, as it would be unfair on specific businesses to expect them to pay themselves. The overall assessment of costs for the country should be balanced by a proper assessment of the costs of *not* going metric. In other words, if the cost were a capital expenditure for a company, in how many years would the benefits exceed the costs.

16.5 Major Problems Due in Part to Mixed Units

Naturally, there is a cost to be borne for changing units as well as long term benefits in using coordinated units. Many of the costs are incalculable or have not been calculated in practice. Nevertheless, the misery and suffering associated with the use of mixed units is an ever present potential danger. I describe three examples which are of a somewhat serious nature. First, I describe an example in which a jet aircraft ran out of fuel. For the second I describe the loss of a NASA satellite. The third also relates to an aircraft accident. The UK Metric Association explains many of the problems arising in general from the use of mixed unit systems[18], while in addition, the US Metric Association provides further specific examples where problems have arisen[19]. Hurrah for the US Metric Association, fighting against tremendous odds!

(i) The Gimli glider
Air Canada Flight 143 on 23 July 1983 became known as the Gimli glider, after the jet aircraft was forced to glide to an emergency landing[20,21]. The Boeing 767 was a relatively new aircraft at this time, and Canada had recently converted to the metric system. A

Measuring the World, by John Austin

near fatal accident occurred after the aircraft ran out of fuel while still some 1½ hours from its destination[20], and cruising at an altitude of 12,000 m. The problems occurred on a flight from Montreal to Edmonton (Canada) with a short stop over in Ottawa. The total flight was to take 4 hours, and to save time, the fuel needed for the full journey was taken on board in Montreal, at least in principle. The Boeing 767 had the facility for carrying up to 50 tonnes of fuel (A-1 kerosene jet fuel). It is powered by two Pratt and Witney fat jet engines consuming 4000 kg of fuel each per hour. Engineers learnt the details of the engine in a mixture of units, with fuel in kilogrammes and thrust in pounds. Mechanics also needed 2 sets of tools (metric and imperial) to work on the aircraft. Due to faulty electronics the fuel gauges were not operating properly, but the aircraft was allowed to fly anyway as a manual test before and after refuelling was carried out to 'confirm' the amount of fuel. There was still 7700 L present in the tanks from the previous flight and was then topped up to provide fuel for the full trip to Edmonton. Calculations indicated that 22,300 kg was needed for the full journey. However, the wrong density (specific gravity) of fuel was assumed when the aircraft was refuelled with a measure in litres. The density assumed was 1.77 kg per litre whereas this is the density in pounds per litre. The proper density is 0.8029 kg L^{-1}, so the aeroplane was actually loaded with just 9,800 kg for a journey requiring at least 20,700 kg (with 22,300 *kg* as the 'target amount'). Discussions between the pilot, co-pilot and the two responsible for refuelling did not expose the problem. Four grown men could not carry out the trivial calculations that should have been easy for any high school student. After the arrival at Ottawa, the fuel level was checked again, and found to be 11,430 L. Again, it was converted to kilogrammes at the wrong density.

After running out of fuel in mid-air, the captain and pilot did a remarkable job in gliding the aircraft to the ground without loss of life. Air Canada later found in simulations that other pilots might have crashed the aeroplane[21]. The loss of the first engine (due to fuel starvation) immediately triggered a 'Major Incident' stimulating the investigation by the Accident Investigation Board. In a 199-page report, published in November 1984, the Captain (Pearson) and co-pilot (Quintal) were commended for their exceptional skill in landing the crippled airliner[22]. Air Canada was largely blamed, which in turn blamed its personnel. Airline deficiencies were considered partly responsible but the report suggested that part of the problem was caused by aircraft specifications based on kilogrammes of fuel, rather than pounds of fuel. In other words, if only 'Air Canada' had resisted government pressure and kept to pounds, the disaster would have been averted. Air Canada itself did punish the Captain and copilot with demotion and temporary release from duty respectively[21]. In practice the fuel conversion was just one of a number of poor communications. To my mind it beggars belief to suggest that we should stick to our ancient units, rather than teach employees to think straight. You could argue, as I do, that while different countries use different units, mistakes of various kinds will come up from time to time. Many businesses of course may well use metric units throughout the business, but mistakes are often picked up by employees with an intuitive grasp of the sizes of quantities. If employees in their normal life use peculiar units then they won't have the intuition to spot a problem and prevent it from becoming serious. How anybody could think that the density of aviation fuel is 1.77 kg L^{-1} is a complete mystery to me, and I would guess that anyone brought up to use the metric system in their normal life would think the same. The cost of repairs to the aeroplane were about $1 million[23].

    This event occurred in the 1980s and the adoption of the metric system and its earlier environmental stance made Canada, to my mind, one of the most admired countries in the world. It has now rather squandered that position by its pursuit of coal tar sands, potentially damaging the environment in the process, and its now incomplete conversion to the metric system.

(ii) Loss of the Mars Climate Observer
One of the most expensive mistakes made with mixed units was made with the Mars Climate Observer, which was heading for the red planet in September 1999[24,25]. A burn on the engine was needed to place the satellite into Mars orbit, but instead of passing 120 km from the surface on closest approach, it passed 57 km from the surface and burned up in the atmosphere Later investigation revealed that the required engine impulse (the product of the thrust and the time over which it operates) was communicated from the engineers, Lockheed Martin, to the JPL navigators in lb s, instead of the N s that were contracted. That is equivalent to giving the engine thrust in pounds instead of newtons, a factor of 4.36 lower (§ 9.3). The financial cost was $125 million, about half the cost of making the UK completely metric, although the mission, which was essentially a write off, actually cost $328 million. But NASA has lots of money, and doesn't need to learn any lessons. They still use Lockheed Martin, who have still not gone metric. Further, as far as I know, Lockheed Martin were not fined for breach of contract. It seems only a matter of time before another spacecraft is lost. Perhaps if a person were on board, it would be a different matter and metrication might have some sort of priority.

(iii) Korean air cargo flight 6316
 A second aviation disaster due to a mix-up over units was the loss of a Korean air cargo plane[26]. The flight took place on 15 April 1999 and crashed soon after leaving Shanghai, bound for Seoul. The captain received clearance to climb to 1500 m altitude, and after reaching almost the correct height, the pilot became confused and thought that he should be at 1500 ft. During the subsequent descent, the pilot lost control of the aircraft which killed all three on board as well as five people on the ground. This

confusion arose from the fact that in the Mid-East/Asian region, air traffic control use metres for altitude[27] rather than feet, as used in other regions. This must be shocking to most people who fly regularly, and is made worse by the fact that nobody uses altitude anyway. Aircraft actually fly along pressure surfaces, with the pressure converted to nominal height. So wouldn't it be simpler to work with the pressure surface itself and be done with it?

16.6 National Efforts Towards Metrication

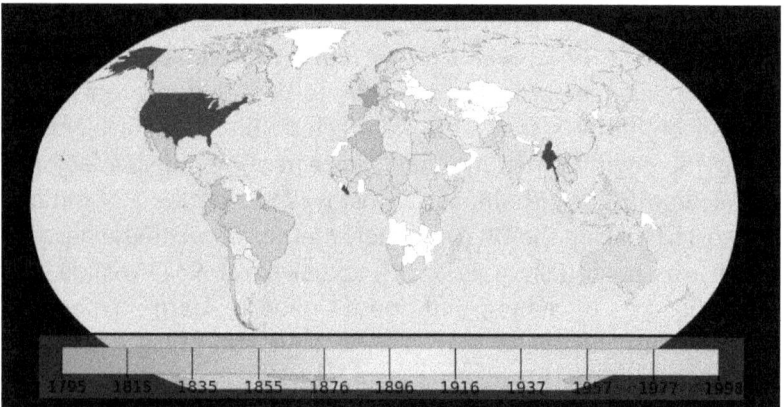

Metrication of the world by date. Image from Wikimedia[29]. After the figure was prepared (31 July 2012), Burma announced its intention to go metric. Note the black area in the North and West of the globe, making the US now a world leader in non-metric units, but nobody seems to be following not even Burma!

Virtually all nations, even the USA, have indicated some desire in using the metric system for its official weights and measures. It is worth while seeing how far the reality has compared with underlying political statements – in many cases not well at all. The reader is directed to the UK Metric Association[2] for its coverage of Australia, Canada, South Africa, New Zealand and Ireland. Australia and Ireland are covered by Wikipedia as already

Measuring the World, by John Austin

referenced[9,10], while a nation by nation summary is given in[28]. The following material is a somewhat subjective assessment of the current situation, based on my personal travels as well as discussions with individuals from the countries themselves. Treat it with a pinch of salt!

(i) UK
As we have seen earlier, the UK has made considerable progress towards metrication, but that effort has stalled before the process was completed. In particular, road signs and speeds are still conveyed in miles and miles per hour, albeit with the disconcerting abbreviation of m for mile. In athletics M has become more common for mile in the few events and records still held over the classical distance. Of course the first sub 4 minute mile by Bannister in 1954 was an important historical, not just sporting moment, and the retention of the mile record maintains the history. Somehow, the recording of the first 4 minute 1500 m does not have the same caché. For women, sub 4 minutes for 1500 m remains a substantial target, but I can't recall the first woman under that barrier. Retention of the mile for certain historical reasons like this adds to the richness of our culture, but the same can't be said of the many other units in existence. I don't think there is a good cultural reason for retaining the mile on British roads. The excuse is that it would cost too much and cause too much chaos to convert all the road signs to kilometres, but if anything the Irish experience shows otherwise.

(ii) Former British Commonwealth
Canada has almost the same problems as the UK in that the metrication process was stopped before completion, apparently for political reasons. In 1975 it agreed to go metric, ten years after Britain. When I've been there, however, metric units were accepted everywhere, like American dollars! Because of its Southern neighbour, though, there is some 'leakage' of units across

the border. Australia and New Zealand were successfully weaned off the Imperial system of units many years ago, having decided to go metric in 1970 and 1969 respectively. I don't have direct experience of other commonwealth countries. I expect that the story is the same in other former commonwealth countries such as India, albeit with perhaps a remaining presence of the old units, after their originators had fled.

(iii) USA
The USA has done least to adopt the metric system but of all nations it can probably benefit the most by it, even if it saves 'only' a few hundred million dollars every now and again by its engineers not making silly mistakes. The amazing thing is that in 1975 (yes, almost 40 years ago), the US passed the Metric Conversion Act, which called for a voluntary change to the metric system. A bill was passed in 1988 declaring that the metric system was "the preferred system of weights and measures for US trade and commerce"[30]. Preferred by whom is not clear, as even government reports such as from the National Oceanic and Atmospheric Administration (NOAA) are written by their own scientists in obscure units. No timetable for the hypothetical 'preference' was specified, nor were there penalties for not converting. One is tempted to think that this bill was a political exercise and there was never any intention to force compliance. In the climate change reports we had to prepare from time to time for NOAA, units were usually required to be in fps, making the results look somewhat unfamiliar to the scientists involved! So much for the Dept. of Commerce (of which NOAA is a part) thinking of metric as the "preferred set of units for commerce"! Some products in the USA are sold in metric-sized packages, but on the whole, the metric system is conspicuous by its absence. Of course history shows that whenever a voluntary change is called upon, nobody does anything. The country should simply not have bothered with the politics. It really shouldn't take much effort to sell Coca Cola in 300 mL cans like the company does

Measuring the World, by John Austin

in other countries, but 12 oz cans (340 mL approx.) persist. The US Metric Association[31] has been attempting for decades to modernise US units, but has to date met with considerable resistance, despite a campaign aimed at business, where the savings are likely to arise. Of course that doesn't stop the US, like the international bully it is, from forcing its rotten units on every one it does business with. About the only thing metric in the USA is a 2 L bottle of soda (which is almost the same size as 2 quarts, so may not have required any equipment changes). The illegal drug cocaine is sold by the gramme, presumably because South America is metric, whereas cannabis is grown locally and sold by the ounce!

(iv) Summary

In the table, I summarise my subjective opinions for just a few countries regarding their use of the metric system.

Region/Country	Score	Comments
Continental Europe	100	Totally metric; fps almost unknown
Asia (Japan, China, Korea)	100	Totally metric; fps almost unknown
Australia/NZ	95	Totally metric; successfully weaned off the fps system, but residuals exists.
Canada	80	Metric in principle, but there is some significant contamination at the border.
Bermuda/Caribbean	75	Officially metric but forced to trade with the USA in funny units.
UK	50	Officially metric, but miles (roads) and stones (people's weight) persist, as do feet and inches and a host of other confusions.
USA	25	Metric behind the scenes in some places, but the metric system is almost entirely unknown by the US public.

Measuring the World, by John Austin

The score (out of 100) is only for entertainment purposes. It would take only an effort by the USA and to a smaller extent Britain and Canada for the whole world to be able to give up on the old unscientific units. Wouldn't that be a global agreement to be proud of, even for the president of the USA?

References

[1] A very British mess, UK Metric Association, http://ukma.org.uk/avbm-summary, accessed 13 April 2014.
[2] A Very British Mess, brochure by the UK Metric Association, 2004, http://ukma.org.uk/sites/default/files/VBM.pdf, accessed 13 April 2014.
[3] Welcome to the British Weights and Measures Association, 2001, http://www.bwmaonline.com/, accessed 13 April 2014.
[4] Imperial language, please, the Economist, 21 March 2012, http://www.economist.com/blogs/johnson/2012/03/weights-and-measures, accessed 13 April 2014.
[5] Marine forecasts glossary, Met. Office, http://www.metoffice.gov.uk/weather/marine/guide/glossary.html, accessed 13 April 2014.
[6] Contrast Britain's decimal currency and metric conversions, UK Metric Association, 2013, http://www.metric.org.uk/decimalisation, accessed 13 April 2014.
[7] History of the Euro, Wikipedia, 27 February 2014, http://en.wikipedia.org/wiki/History_of_the_euro, accessed 13 April 2014.
[8] Metrication in the United Kingdom, Wikipedia, 12 April 2014, http://en.wikipedia.org/wiki/Metrication_in_the_United_Kingdom, accessed 13 April 2014.
[9] Metrication in Australia, Wikipedia, 28 February 2014, http://en.wikipedia.org/wiki/Metrication_in_Australia, accessed 13 April 2014.
[10] Metrication in Ireland, Wikipedia, 14 January 2014, http://en.wikipedia.org/wiki/Metrication_in_Ireland, accessed 13 April 2014.
[11] The world is flat, A brief history of the Twenty-first Century, Thomas.

Measuring the World, by John Austin

L. Freeman, Farrar, Straus & Groux(Pub.), April 20005, http://www.thomaslfriedman.com/bookshelf/the-world-is-flat.
[12] Road speed limits in the Republic of Ireland, Wikipedia, 3 June 2013, http://en.wikipedia.org/wiki/Road_speed_limits_in_the_Republic_of_Ireland, accessed 14 April 2014.
[13] Metric signs ahead, UK Metric Association, 2009, http://ukma.org.uk/sites/default/files/MSA.pdf, accessed 14 April 2014.
[14] Are motorists overtaxed?, Taxes and charges on road users, Transport Committee, 24 July 2009, http://www.publications.parliament.uk/pa/cm200809/cmselect/cmtran/103/10304.htm, accessed 14 April 2014.
[15] Estimating the cost of conversion of road traffic signs to metric units, Department for Transport, http://webarchive.nationalarchives.gov.uk/+/http://www2.dft.gov.uk/pgr/roads/tss/gpg/estimatingcostconversion.html, accessed 14 April 2014.
[16] Historic inflation calculator: how the money has changed since 1900, R. Browning, This is money.co.uk, http://www.thisismoney.co.uk/money/bills/article-1633409/Historic-inflation-calculator-value-money-changed-1900.html, accessed 14 April 2014.
[17] British traditions at risk from Sunday roast to doorstep pint of milk, R. McPhee, Mirror Newspapers, 1 October 2013, http://www.mirror.co.uk/news/uk-news/british-traditions-risk-sunday-roasts-2325406, accessed 14 April 2014.
[18] Problems arising from two systems, UK Metric Association, 2013, http://www.metric.org.uk/problems-arising-from-two-systems, accessed 14 April 2014.
[19] Unit mixups, US Metric Association, http://lamar.colostate.edu/~hillger/unit-mixups.html, accessed 14 April 2014.
[20] "Freefall", William and Marilyn Hoffer, Grafton Books, 1989.
[21] Gimli glider, Wikipedia, 25 March 2014, http://en.wikipedia.org/wiki/Gimli_Glider, accessed 14 April 2014.
[22] Final report of the Board of Inquiry investigating the circumstances of an accident involving the Air Canada Boeing 767 aircraft C-GAUN that effected an emergency landing at Gimli, Manitoba on the 23rd day of July, *1983,* Commissioner, G. H. Lockwood, Government of Canada, Ottawa,

1985.

[23] Gimli glider July 23rd, 1983, Gimli, Manitoba, http://www.gimlicommunityweb.com/history/gimli_glider.php, accessed 14 April 2014.

[24] Mars climate orbiter, Wikipedia, 3 April 2014, http://en.wikipedia.org/wiki/Mars_Climate_Orbiter, accessed 14 April 2014.

[25] "Roving Mars: Spirit, Opportunity and the exploration of the red planet", Steve Squyres, Hyperion Books, New York, New York, 2005.

[26] Korean air cargo flight 6316, Wikipedia, 10 March 2014, http://en.wikipedia.org/wiki/Korean_Air_Cargo_Flight_6316, accessed 14 April 2014.

[27] International operations manual, Region: Middle East/Asia, 12 January 2014, http://code7700.com/iom_midasia.html, accessed 14 April 2014.

[28] Metrication, Wikipedia, 14 April 2014, http://en.wikipedia.org/wiki/Metrication, accessed 15 April 2014.

[29] Metrication of the world by year, image from Wikimedia, 31 July 2012, http://en.wikipedia.org/wiki/File:Metrication_by_year_map.svg, accessed 15 April 2014.

[30] Metrication in the United states, Wikipedia, 11 April 2014. http://en.wikipedia.org/wiki/Metrication_in_the_United_States, 15 April 2014.

[31] US Metric Association, http://lamar.colostate.edu/~hillger/, accessed 15 April 2014.

Measuring the World, by John Austin

17. The Future of the Metric System

17.1 Introduction

We have now reached an exciting time regarding for the metric system! The international prototype kilogramme is almost ready to be abandoned. In other words, fundamental experiments now exist which can compare forces, and relate them to the kilogramme approaching the precision that individual masses can be compared with the prototype kilogramme[1]. Once further improvements in experiments occur, the transformation can take place, perhaps as soon as 2015. Our unit system will then be universal: capable of being reproduced at any place in the universe (literally). Fps units remain, as before, defined in terms of SI units, 1 ft = 0.3048 m (exactly) etc.

The SI system relies on the kilogramme for a number of its subsidiary units. Once it is redesigned, a radical transformation of the system will take place. Although the changes are a matter of philosophy in some respects, fixing the values of the fundamental constants would have the effect of fixing our unit system, i.e. the sizes of the units we use. This would be to the benefit of future technology. For example, when the metre was redefined in terms of the speed of light, it immediately resulted in higher precision length measurements to be made which ultimately improved the quality of gps tracking. At the moment, there is a problem with the prototype kilogramme as comparisons have shown that its mass may be drifting[2], but it is unclear what the magnitude of the drift is, and the causes of it. In § 17.2 these problems are described, and the experiments made to resolve it are explained.

As described in § 17.3, the experiments lead naturally to a revision of the SI system by specifying a minimum set of fundamental physical constants. This defines all our unit sizes precisely and all other measurements are necessarily derivable from the base units obtained. The conditions needed for the changeover to the new system are described in § 17.4.

17.2 The kilogramme

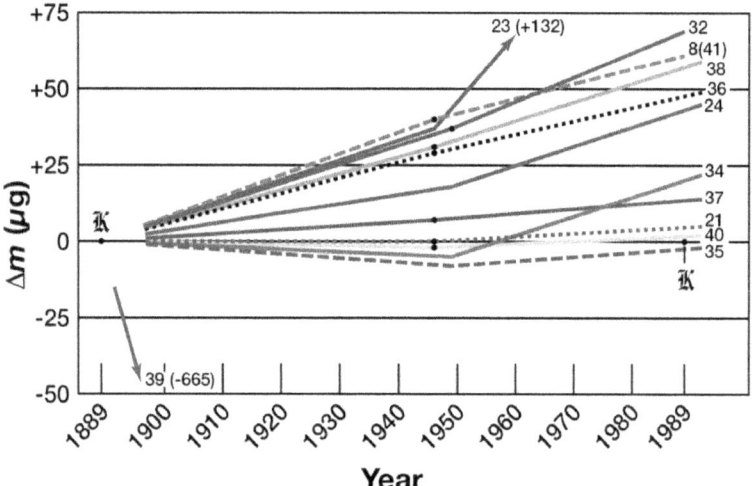

Graph of the relative change in mass of selected kilogramme prototypes, taken from Wikimedia[3]. Each line indicates the mass of each separate national standard relative to the IPK, which is assumed constant. However, there is no way of knowing for certain whether the IPK itself has gained or lost mass.

When the international prototype kilogramme (IPK) was engineered, six additional copies were made, and these have been compared with the IPK on three occasions(figure). The results

Measuring the World, by John Austin

indicate an upwards drift in the copies relative to the IPK[2,3,4] and hence the plan to change the SI units was instigated to ensure the long-term stability of the mass unit. The typical drift has been about 50 μg per century, about 5 parts in 10^8 in total. Two experiments have linked the Planck constant to the mass of the IPK with an uncertainty of less than 5 parts in 10^8, and these experiments are the focus of improved specifications of the SI units.

In the process of revising the SI units, the Planck constant (Chapter 2), h, the fundamental constant which relates energy to frequency in an atomic oscillator[5] will be measured by two separate experiments. The experiments are linked through the molar Planck constant $N_A h$ which is very well known[6], through the Rydberg constant, from spectroscopic theory and measurements[7]. In other words,

$$N_A h \; (= 3.9903127176(28) \times 10^{-10}) \text{ J s mol}^{-1}$$

where the number 28 in brackets is the uncertainty, corresponding to only 0.7 parts per billion (ppb), much smaller than a direct measurement of the IPK (about 20 ppb). So, essentially, in the current SI a measurement of N_A is equivalent to a measurement of h and the converse. Once consistency is reached in the values of h, the kilogramme can be redefined in terms of the Planck constant (§ 17.3).

(i) Experiment 1.

The International Avogadro Coordination, now the International Avogadro Project (IAP)[8], aims to measure the mass of the silicon-28 atom, m_{Si}, relative to the mass of the IPK, to determine the Avogadro constant, N_A. Silicon is used, because it can be grown as highly pure, almost perfect crystals. As shown in § 13.4, the value of N_A is determined from the density of the silicon crystals, the spacing of the atoms within the crystal, and the atomic mass of the silicon atom. The best result so far obtained[9], in terms of the highest precision, has a measurement uncertainty of 30 ppb:

$$N_A = 6.02214082(18) \times 10^{23} \text{ mol}^{-1}$$

or in terms of h, $h = 3.9903127176(28) \times 10^{-10}/N_A$ giving

$$h = 6.62607009(20) \times 10^{-34} \text{ J s}$$

in all these numbers, the digits in brackets represent the uncertainty in the last two digits. This experiment although simple to understand in principle is technologically very challenging because of the need firstly to obtain a pure crystal of silicon and then to engineer it to the correct dimensions for a 1 kg sphere. Measuring the volume of the sphere was the largest uncertainty of the experiment, contributing over half the total uncertainty. The technique used was optical interferometer which was able to measure the radius of the sphere to approximately the nearest layer of atoms, 0.3 nm[4].

(ii) Experiment 2.

The watt balance measures the Planck constant relative to the mass of the IPK[10,11]. It consists of two subsidiary experiments in which a test mass and coil are suspended from a balance in a magnetic field. In the first experiment, the coil is stationary and the gravitational force is balanced by a current through the coil. In the second experiment the coil is moved at a known speed. The characteristics (wire length and magnetic field) can be eliminated from the equations, and the power expended equals the work done in supporting the test mass. The power consumed by the watt balance can be described in terms of the Planck constant, which enables the test mass and hence the IPK to be determined in terms of h. The watt balance is at the absolute limit of technology. It takes about 10 years to build the apparatus from scratch[11] and so relatively few countries have them. The most precise value so far obtained is with the NIST Watt balance, with an uncertainty of 36 ppb in 2007. More recent measurements at NIST[12] have been slightly lower, but with a larger uncertainty of 45 ppb. Nonetheless, the experimenters believe that this more recent value is superior due to a complete readjustment of the apparatus and remeasurement of all key parameters. The value obtained for Planck's constant was

$$h = 6.62606979(30) \times 10^{-34} \text{ J s}$$

This value was obtained with the instrument NIST-3, and the NIST-4 Watt balance is being assembled to replace it. Another reason to prefer this result over the earlier result is that it differs from the IAP value by 30 in the last two digits, precisely the size of the uncertainty. In other words, both new measurements in h suggest that an upward adjustment is needed and the two values have come into agreement, essential for the revision of the SI (§ 17.4). Make no mistake, these are remarkably high quality experiments performed at the leading edge of science and technology. US scientists are making world leading contributions to the SI system and it's quite ironic and disappointing that their own country seems to be ignoring the effort!

17.3 The Fundamental constants of the SI System

In Chapter 2, we showed how in a few instances the units can be redefined in terms of fundamental physical constants which are then universal, so our putative alien in a distant galaxy could reproduce accurately. The examples given were impractical for every day use because they relate rather to the high energy sub-atomic scale. Instead, the same principle could be applied to the units that we do have so that their sizes do not change but instead their definitions change. In the new SI units[13], the plan is to specify the following constants, most of which appeared in § 2.6, as observed quantities. Currently, apart from the number of wavelengths of caesium to define the second, and the speed of light, all the other constants are measured in the laboratory. The small set of constants, just 7, is all that we need to define the SI units immutably. Once the agreement takes place, the units cannot change, and the drifts in the prototype kilogrammes shown in § 17.2 cannot occur.

(i) The constants
The constants below are based on the 2010 Codata adjustment[14], although precise values will be subject to modification.

1. The unperturbed ground state hyperfine splitting frequency of the caesium 133 atom is exactly **9 192 631 770 Hz**.

2. The speed of light in vacuum c is exactly **299 792 458 m s^{-1}**.

3. The Planck constant h is exactly **6.626 069 57×10^{-34} J s**.

4. The elementary charge e is exactly **1.602 176 565 ×10^{-19} C**.

5. The Boltzmann constant k is exactly **1.380 648 8 ×10^{-23} J K^{-1}**.

6. The Avogadro constant N_A is exactly **6.022 141 29 ×10^{23} mol^{-1}**.

7. The luminous efficacy K_{cd} of monochromatic radiation of frequency 540 ×10^{12} Hz is exactly **683 lm W^{-1}**

(ii) Recovery of units from the fundamental constants
These constants are not obviously the same as our current units. In other words, there is no second, kilogramme or kelvin constant! This probably reflects that human experience on which our units are based is not fundamental in the sense that physics would recognise. Nonetheless, recovery of the SI units is straightforward.

Measuring the World, by John Austin

The Watt balance at the US National Institute of Standards and Technology[15]. This instrument, or a version of it, is proposed as the new mass determining instrument in the new SI units.

The first constant defines the length of the second as in the current SI unit. This definition of the second has been used since 1967 and has been discussed briefly in § 3.8.

The second constant, c, defines the length scale of the metre, which translates as the distance travelled in 1/299,792,458 seconds. Again, this constant has been in use since 1983 and has been described in § 5.3.

The third constant, h, the Planck constant, will be used

to redefine the kilogramme making the prototype unnecessary. Masses will still be compared as now, but whenever precise comparisons or determinations of mass are needed, they will be completed with the Watt balance as explained in § 17.2. The experiment will enable masses to be determined from the fundamental constants, which will be specified as exact numbers similar to the above constants. The changes proposed for the kilogramme would be the most radical of all the unit changes.

The fourth constant, e, redefines the ampere as the rate of flow of charge. The definition of the ampere is based on the force between two current carrying conductors and has the effect of specifying the magnetic constant of the vacuum, μ_o to be exactly $4\pi \times 10^{-7}$ H m^{-1}. This constant, as well as its counterpart the electric constant ε_o, relate magnetic and electric fields in Maxwell's equations of electromagnetism[16]. They are still related by $\mu_o \varepsilon_o = 1/c^2$, which of course is specified in SI. However, with the new definition, μ_o and ε_o need to be experimentally determined, although the relative uncertainty is less than 1 part in 10^9.

The fifth constant, k, the Boltzmann constant, redefines the kelvin in terms of thermodynamic energy. Currently, the kelvin is defined by specifying the exact temperature of the triple point of water (273.16), combined with the temperature at absolute zero (0 K). In the new system, the triple point will be a measurable quantity, and the expected uncertainty based on the specification on the Boltzmann constant is about 3×10^{-4} K. One of the difficulties of the current standard is the problem of obtaining a sufficiently pure source of water.

The sixth constant, N_A, the Avogadro constant, defines the size of the mole. Whereas previously, N_A was experimentally determined from the definition of the mole -- the number of atoms

Measuring the World, by John Austin

in 12 g of Carbon-12 -- the definition is now turned around, so that the Avogadro constant is specified and now the number of atoms in 12 g of Carbon-12 is to be determined experimentally. With the new definition, that number will be 1 mole to a precision of 1 part in 10^9. with the proposed changes, the mole will no longer be dependent on the definition of the kilogramme. This is an important change, as the previous connection to mass obscured the fundamental nature of the constant.

The seventh constant, K_{cd}, is a reworking of the definition of the lumen discussed in Chapter 14. There, the definition related to the candela which is related to the lumen by summing up over all solid angles. In other words the light output in lumen is the luminance in candela x the solid angle over which the light is observed.

Given that the final decision on the numerical values of the above constants will determine our unit system, it is important to have the best known values for them all. Many of the constants have been set with N_A and h being the two critical ones as noted above. The remaining quantities to be remeasured are the charge on the electron and the Boltzmann constant. The latest measurement of the latter[17] by the UK National Physics Laboratory, has reduced the uncertainty from 0.94 parts per million (ppm) to 0.71 ppm to arrive at a value

$$k = 1.38065156(98) \times 10^{-23} \text{ J K}^{-1}$$

This is slightly higher than the CODATA value in the above set of constants, but consistent with the uncertainty analysis.

Improvements in the measurements of the charge on the electron are not a priority, as the value is already known to a precision of 20 ppb.

17.4 Adoption of the new units

The above ideas are in principle just a proposal and may not be realised in practice. Conditions for it to happen are that the uncertainties in the measurements of h need to drop to 20 ppb[8] and there needs to be three measurements within 50 ppb. Even if both these conditions are met, there is no guarantee that the change will go ahead. Several years ago it was starting to appear problematic as a 2011 measurement of N_A by the International Avogadro Project (IAP)[9] yielded a value which differs from the official CODATA 2010 value by about 8 parts in 10^8, twice the quoted CODATA uncertainty. While such a statistical fluctuation is not impossible, it should only occur with 5% probablility. This suggested that further measurements of h were needed, which might have delayed the adoption of the new units. However, with the recent measurements from NIST[12] confirming a slightly higher value of h, a result in excellent agreement with the IAP measurement, the plans appear to be back on track.

Once the changes take place, it is the magnitude of our unit system which will be fixed. The fundamental constants themselves are given by nature, but what we can do is produce an independent set of constants which together determine the size of the SI units themselves. By choosing the values which are experimentally determined as the fixed constants, the adjustment to the unit values will be negligible, but provide significant advantages for technical work. In particular, looking at the fundamental constants in § 2.6, with the new SI units, all of the constants given in the table will be exact in SI units, or reduced significantly, except for G which will be unchanged.

Measuring the World, by John Austin

References

[1] Watt balance, Wikipedia, 31 January 2014, http://en.wikipedia.org/wiki/Watt_balance, accessed 15 April 2014.

[2] Towards a definition of the kilogram, BIPM, Sèvres, France, http://www.bipm.org/en/scientific/elec/watt_balance/, accessed 15 April 2014.

[3] Prototype mass drifts, Wikimedia, from The third periodic verification of national prototypes of the kilogram (1988-1992), G. Girard, Metrologia, 31, 317-336, 1994. Attribution: Greg L at the English Language Wikipedia, http://en.wikipedia.org/wiki/File:Prototype_mass_drifts.jpg, accessed 15 April 2014.

[4] Kilogram, Wikipedia, 3 April 2014, http://en.wikipedia.org/wiki/Kilogram, accessed 15 April 2014.

[5] Planck constant, Wikipedia, 8 April 2014, http://en.wikipedia.org/wiki/Planck_constant, accessed 16 April 2014.

[6] Physical constant, Wikipedia, 13 February 2014, http://en.wikipedia.org/wiki/Physical_constant, accessed 16 April 2014.

[7] Becker, P. and H. Bettin, 2011: The Avogadro constant: determining the number of atoms in a single-crystal ^{28}Si sphere, Phil. Trans. R. Soc. A, 2011 369, doi: 10.1098/rsta.0222, 19 September 2011, http://rsta.royalsocietypublishing.org/content/369/1953/3925.full.pdf, accessed 15 April 2014.

[8] International Avogadro Project, BIPM, http://www.bipm.org/en/scientific/mass/avogadro/, accessed 15 April 2015.

[9] Andreas, B. et al., 2011: An accurate determination of the Avogadro constant by counting the atoms in a 28Si crystal, Phys. Rev. Lett., 106, 030801, doi: 10.1103/PhysRevLett.106.030801, 18 January 2011, http://arxiv.org/abs/1010.2317, accessed 4 April 2014.

[10] The principle of the watt balance, BIPM, Sèvres, France, http://www.bipm.org/en/scientific/elec/watt_balance/wb_principle.html, accessed 15 April 2014.

[11]Watt balances, National Physics Laboratory, 2 May 2013, http://www.npl.co.uk/engineering-measurements/mass-force-pressure/mass/research/npl-watt-balance, accessed 15 April 2014.

[12] New value for the Planck constant may hasten electronic kilogram,

NIST Physical Measurement Laboratory, 14 January 2014, http://www.nist.gov/pml/div684/planck_constant_value.cfm, accessed 16 April 20014.

[13] 9th SI Brochure, 16 Dec. 2013 draft, published by the Bureau International des Poids at Mesures, Sèvres, France, http://www.bipm.org/utils/common/pdf/si_brochure_draft_ch123.pdf, accessed 16 April 2014.

[14] Mohr, P. et al., 2012: CODATA recommended values of the fundamental physical constants: 2010, Reviews of Modern Physics, Vol. 84, p.1528-1605.

[15] Watt balance, Wikimedia, Photo by Richard Steiner uploaded Greg L at en.wikipedia, 17 October 2007, http://commons.wikimedia.org/wiki/File:Watt_balance,_large_view.jpg, accessed 16 April 2014.

[16] Maxwell's equations, Wikipedia, 8 April 2014, http://en.wikipedia.org/wiki/Maxwell's_equations, accessed 16 April 2014.

[17] Most accurate measurement of Boltzmann constant yet, Phys.org, 10 July 2013, http://phys.org/news/2013-07-accurate-boltzmann-constant.html, accessed 26 April 2014.

Measuring the World, by John Austin

18. Future Unit Use

The design of unit systems is, arguably, one of mankind's major intellectual achievements. While the achievements have not captured the public imagination to the extent that, say, Einstein's contributions have, they are every bit as useful. The design has evolved over the last century or more, largely absent from the public consciousness but hopefully, this book will have gone some way towards conveying those scientific and technical achievements in the context of our every day lives. The information is in the public domain of course, but most of it is not generally needed for practical purposes. That has been perhaps one of the greatest achievements in the field: the transformation of the underlying basis of the units without the user needing to known every detail. Think of the metre: originally one ten millionth the distance from the pole to the Equator but by 1983 it had quietly been replaced by the distance travelled by light in a fraction of a second, having acquired other definitions in between. At no stage has the metre changed by a measurable amount, but its adjustment over the decades has reflected improved technology and the ability and desire to make measurements to higher and higher precision. It's still the same metre, but it's now on a robust footing in a logically and technologically consistent framework.

I remain of course continually frustrated by the fps system, also known as Imperial units, or US customary units. They are a hotchpotch collection of units cobbled together over centuries and shoehorned into the changing engineering situation at the time. They vary according to which side of the Atlantic Ocean you happen to live. They are unnecessarily complicated, including features which are not consistent with physics, such as differences between liquid and dry measures, and differences between heat, dietary and electrical energies. They are unable to cope with very large or very small measurements. Finally, for modern applications (e.g. electricity), the fps doesn't have any units of its own so it tends to borrow from the metric system. This makes it all very difficult to convert from one form of energy to another, for example, even to find out whether it's best to heat your home with gas or electricity. All of these issues have been raised in this book at some point. Often, fps units are close to some metric unit anyway. For example, most people can't judge distance to better than 25%, at best. So why not just abandon the yard, and think of the distance in yards as the same distance in metres instead. That would only lead to a 10% error anyway, and earlier in the book I have offered better approximations which can be computed mentally if needed. The problem is that while people call a certain distance a yard, it has to be a yard, which has its own definition. This leads to a waste of resources in double labelling of critical distances, such as heights of bridges on British roads as well as a host of other issues.

Of course a major step forward has already been taken to make the Imperial set of units entirely dependent on the metric system. In other words, the inch is defined as 2.54 cm exactly and the same applies to other fps related units. There are different sets of units on both sides of the Atlantic, but the US has its own numbers in relation to metric units and the rest of the world (those daft enough to bother with fps) has a different set of numbers. Science itself has long since abandoned non-metric units. UK and US scientists have been honoured in having metric units named after them and other UK and US scientists are currently making world leading contributions to the metric system. For

Measuring the World, by John Austin

example, in making the exquisite measurements which will probably shortly become the new mass standard to replace the kilogramme. It is strange, and I believe reflects badly on the profession, that engineering persists with fps units, especially in the USA. You would have thought that somebody might have learnt something from the loss of an expensive satellite.

I end this book with some pleas.

(i) US engineers

My first plea is to US engineers to listen to your own scientists, stop making excuses about the cost, and convert forthwith from fps to metric units. We know it's not about the cost, it's just ideology. Get over it! Democracy doesn't always have the right answer, but 90% of the world surely can't be wrong. Once the changeover happens, engineers can be confident that the unit system is rigorously established at the highest levels of precision, and that they are unlikely to lose any more satellites with silly mistakes.

(ii) International business

My second plea is in the area of international business. In my opinion there are still too many funny units around. In your business, you may know what a barrel of oil is, or a bushel of corn, but do your national and international customers? There may even be chief executives of oil companies as well that don't know how many litres of oil are in a barrel. Are there advantages in the public knowing what these units mean, or do you prefer to keep them in the dark? I know its a different type of corn from sweet corn, but the wholesale price of corn varies substantially according to weather conditions. If the price is $7 per bushel, is that high or low compared with supermarket prices? That actually works out at say 20c per litre which still seems cheap, even during a drought, perhaps a factor of 5 or 6 below the retail price. However, to do the

Measuring the World, by John Austin

calculation you need to know what a bushel is. As noted before, most people I've met don't know whether it's a volume or a mass, let alone its magnitude. Often the corn price is a lot lower than $7 per bushel. I think the public deserves to know what the real price is, and where the markup to retail prices is taking place. By contrast, the markup in the USA on gasoline seems relatively small compared with the barrel price (60c per litre, based on a $96 barrel). This works out at $2.27 per gallon, compared with the retail price of about $3.60). Why isn't the oil industry making that clear by transparent pricing? Transparent pricing would also help to sniff out, for example, UK government hypocrisy. For example, if we know the wholesale price of oil in proper units, we can see where the UK government is levying taxes. So, when it publicly complains about oil companies profiteering, it becomes a little more obvious that the main organisation profiting from oil prices is the UK government itself with its high fuel taxes. Far be it for me to be an apologist for the oil companies, I just seek open and balanced reporting.

When you read the data from the financial markets you get a cornucopia of exotic and modern units. This just reflects the nature of the underlying industry. But it might be useful to be able to compare easily the price of copper for example (usually quoted in $ per tonne) with the price of gold in (in $ per tr oz, whatever that is, but see Chapter 7!). The gold price is usually quoted as a price per ounce (as a shorthand for troy ounce), and I expect many people think (wrongly) that they know what that is. Then for reasons I don't entirely understand, short tons (2000 lb) are often used instead of long tons (2240 lb), but without full explanation. So, in practice I'm left wondering how many pounds I'm working with, and it is usually only the metric conversion that gives me a hint. I think the relevant US businesses just can't be bothered to divide by a peculiar number (2240), which is fair enough. But, then, that was one of the reasons why the metric system was designed. So my plea to international business and finance is to use the metric system, which was designed for your benefit, so that transactions are transparent.

(iii) Governments

My third plea is directed at governments, particularly the UK and USA, to follow the lead of Australia and other Commonwealth countries, away from fps units. My plea is that all governments should actually use the metric system, and not just 'prefer' it, like the USA claims. Even the UK, which is arguably one of the most mixed-up countries still, we are officially metric. However, old units abound. The EEC could actually do something useful and at least get us to make up our minds whether we are metric or fps to an agreed timetable. Perhaps it would be prepared to accept the cost of changing all our road signs, if that's what it takes. In the meantime, it appears that the UK government has decided the short-term cost of each change and implemented only the cheapest on a piecemeal basis with the result that we are neither metric nor fps. At least people are not mystified when you quote a distance in metres. No doubt democracy plays a role, but leadership counts as well, and while the UK leadership appears not to be committed to the metric system (as indicated by all the road signs in the country), the people won't make an effort either. So my plea to the UK government is to change all the road signs, over a committed period of time, to km and km/h, use proper metric notation on the roads (no kph please or m to mean miles), and to insist that all new cars are instrumented only in metric, also from a certain date onwards. There is plenty of evidence, from Ireland for example, which suggests that the speed limit conversion can take place over a weekend without major problems.

In the USA it almost seems that fps units are part of the US constitution, and the US constitution is like a set of commandments from God, which results in murdered schoolchildren while upholding the rights of citizens to bear arms. To change the constitution and by a *reductio ad absurdum*, to change the unit system would be considered 'unpatriotic'. Instead, metric units are often described as 'scientific units', which annoys me. I don't suppose that your average Mexican citizen seeing 30° in the weather forecast thinks he is being 'scientific' when he realises it's going to be a warm day. No science is required. The problem with labelling the units as 'scientific' is that it creates barriers. Because of the poor education system in a lot of developed countries, like the USA, science and mathematics are considered 'hard', so anything labelled 'scientific' is generally avoided. Of course if you are living in a remote community using fps or by inference 'unscientific' units, there is no problem. Difficulties arise only when you leave, such as if you become a tourist in another country.

Since it will be a long time before a 'full plea' to the US government to adopt metric units could bear fruit, I could offer a few mini-goals along the way to metrication. If the units themselves can't be immediately adopted, at least the spellings should be to avoid undermining global agreements. And stop calling them *metric tons:* the name of the unit is *tonne*. Even in the 25th century Americans will still be calling them metric tons, if 'Star trek' is an accurate predictor, and Patrick Stewart is British! A second goal would be to improve the education system so that metric units are better understood. A third goal would be to stop calling them 'scientific units', for the reasons noted above: they should be called 'metric' units.

(iv) The public
My final plea is perhaps more for the British public than any other citizens. For those people who still use imperial units, you may not understand them as well as you think. And why are you mixing fps and metric units all the time in your daily life? Isn't this difficult? So, consider the metric system as an alternative. If you're younger than me (which wouldn't be difficult), at least your education has

covered them, unlike probably the fps system. At the end of this book, I give conversion tables, but try not to use them much. The simplest way to change is to go 'cold turkey'. I've always been in the situation where I have been able to do the arithmetic mentally to go from one system to the next, but one can readily get used to metric units. They are, after all, designed to be a practical size. As in the fps, we can't expect everything to come out how we would ideally like it. Let's just consider temperature. A low temperature of $32^{\circ}F$ isn't emotive, if you're not used to it, but $0^{\circ}C$ might seem so much colder. A high temperature of $100^{\circ}F$ is emotive, but then so is $40^{\circ}C$ once you're used to it. It's all very subjective but it is easy enough to switch. I've thought in metric for a long time now, and found it much easier to work with the measurements. If you're having difficulties working out how many acres are in a certain land area why not give metric units a try? Surely, you owe it to yourself to know your weight and height in kilogrammes and metres so that you can calculate your own body mass index, even if it's just to check on your doctor's numbers.

To me the metric system, especially with the proposals described in Chapter 17, is like a symphony by a famous composer. The composer is not Mozart, whose music is perhaps incomplete. Rather, the composer is like Mahler or Shostakovich, men who have stood on the shoulders of giants to quote Newton. The music has been brilliantly composed and may yet evolve further. Even if the music doesn't change much beyond that outlined in Chapter 17, the public and industry will likely benefit in ways beyond that currently recognised. I hope this book has improved that recognition. Many people in their lifetime would like to see a person land on Mars, or an end to cancer, or something equally profound. I have the rather sad aim of finally seeing the end of those awful units except in the history books. You know what units I mean.

Measuring the World, by John Austin

Acknowledgements

Alan Hesketh provided a valuable sounding board for much of the material of this book during its formative months. The US Government has been a good sport in unwittingly taking my numerous pot shots on its collective chin. I do genuinely thank the US Government for not copyrighting its material, making life a lot easier to incorporate figures into this book, and indeed in my other science articles. Next time, though, can I ask you to remove the funny units from the diagrams? Wikipedia provided valuable information for the research and fact checking stages and 10% of the profit from book sales will be donated to Wikipedia in acknowledgement of its assistance. A further 10% of the profit from sales will be donated to the British Metric Association to further the cause of British metrication, started a mere 50 years ago!

Measuring the World, by John Austin

Appendix A: Definition of terms and derived units for the metric system

A.1 Metric Multipliers

10^{-24}	yocto, y	10^{-3}	milli, m	10^6	Mega, M
10^{-21}	zepto, z	10^{-2}	centi, c	10^9	Giga, G
10^{-18}	atto, a	10^{-1}	deci, d	10^{12}	Terra, T
10^{-15}	femto, f	1	-	10^{15}	Peta, P
10^{-12}	pico, p	10	deca, da	10^{18}	Exa, E
10^{-9}	nano, n	10^2	hecto, h	10^{21}	Zetta, Z
10^{-6}	micro, µ	10^3	kilo, k	10^{24}	Yotta, Y

Usage: mm = 10^{-3} m, km = 10^3 m,....., Em = 10^{18} m etc.

A.2 Definition of the SI base units

second (s): 9 192 631 770 periods of the electron transition between two hyperfine ground state levels of Caesium-133.
metre (m): The distance travelled by light in a vacuum in 1/299,792,458 seconds.
kilogramme (kg): The mass of the prototype kilogramme held at the institute of weights and measures in Sèvres, France.
kelvin (K): 1/273.16 of the thermodynamic temperature change between absolute zero and the triple point of water.
ampere (A): That current, which if flowing in two wires a metre apart in a vacuum, produces a force of 2×10^{-7} newtons per metre of length.
mole (mol): The amount of substance containing the same numbers of atoms or molecules as 12 g of Carbon-12.

Measuring the World, by John Austin

candela (cd): The light energy emitted per solid angle at a frequency 5.4×10^{14} Hz in units of 1/683 W per steradian.

A.3 List of SI units

Base Units

Unit	Abbreviation	Use	Main Chapter Reference
second	s	Time	3.8
metre	m	Length	5.3
kilogramme	kg	Mass	7.2
kelvin	K	Temperature	8.3
ampere	A	Electrical current	12.2
mole	mol	Amount of substance	13.1
candela	cd	Luminous intensity	14.6

Derived Units

Unit	Abbreviation	Use	Main Chapter Reference
hertz	$Hz = s^{-1}$	frequency	4.2
becquerel	$Bq = s^{-1}$	radioactivity	4.3
gray	$Gy = J\,kg^{-1}$	Radiation exposure	4.4
sievert	$Sv = J kg^{-1}$	Human radiation exposure	4.4
Celsius	$^0C = K - 273.15$	Temperature	8.2

Measuring the World, by John Austin

Unit	Abbreviation	Use	Main Chapter Reference
newton	$N = kg\ m\ s^{-2}$	Force	9.2
radian	$rad = m\ m^{-1}$	Angle	9.11
pascal	$Pa = N\ m^{-2}$	Pressure	10.2
joule	$J = N\ m$	Energy	11.2
watt	$W = J\ s^{-1}$	Power	11.5
coulomb	$C = A\ s$	Electrical charge	12.2
volt	$V = W\ A^{-1}$	Electromotive force	12.2
ohm	$\Omega = W\ A^{-2}$	Electrical resistance	12.2
siemens	$S = \Omega^{-1}$	Electrical conductance	12.2
farad	$F = C\ V^{-1}$	Electrical capacitance	12.3
henry	$H = V\ s\ A^{-1}$	Electrical inductance	12.4
tesla	$T = N\ s\ C^{-1}\ m^{-1}$	Magnetic field strength	12.6
weber	$Wb = T\ m^2$	Magnetic flux	12.6
katal	$kat = mol\ s^{-1}$	Catalytic activity	13.5
lumen	$lm = cd\ sterad$	Luminous flux	14.6
lux	$lx = lm\ m^{-2}$	Light brightness	14.6
steradian	$sterad = m^2\ m^{-2}$	Solid angle	14.6

Measuring the World, by John Austin

A.4 Definition of common terms with SI unit

Acceleration (m s^{-2}): Rate of change of *velocity*.
Area (m^2): A measure of a planar region or the surface of a solid.
Capacitance (F): Property of a material or object which allows it to store *charge*.
Charge (C): That which produces an electric field and opposes or attracts other charges.
Current (A): Rate of flow of electrical *charge*.
Density (kg m^{-3}): Mass of a substance per unit *volume* of space.
Energy (J): A quantity capable of doing *work*.
Force (N): That which causes a mass to *accelerate*.
Frequency (Hz): The number of oscillations or changes per unit time.
Length (m): The minimum distance between two points.
Luminous flux (lm): Light intensity of an object summed in all directions weighted by the perception of the human eye.
Luminosity (W): The total brightness of an object summed in all directions.
Mass (kg): Amount of substance; the measure of a body's resistance to acceleration.
Mole (mol): The number of atoms or molecules in a mass of substance relative to the number of atoms in 12 g of Carbon-12.
Power (W): The rate of doing work.
Pressure (Pa): The *force* per unit *area* acting on a surface.
Resistance (Ω): Property of a material which absorbs an electrical *current*.
Specific Gravity (1): Ratio of the *density* of a liquid to that of water at 4°C.
Speed (m s^{-1}): The magnitude of the *velocity*.
Surface Tension (N m^{-1}): The *force* per unit *length* along a line drawn in a fluid.
Time (s): Elapsed period between one historical point and another

or between any two points in the past or future.
Velocity (m s⁻¹): Rate of change of position.
Viscosity (Pa s): The fluid equivalent of friction: a *force* which opposes the movement of fluid layers.
Voltage (V): Electromotive *force* driving an electrical circuit.
Volume (m³): The size of a three-dimensional object or space.
Weight (N): The *force* of a *mass* under gravity, often used to refer to the *mass* itself (units of kg).
Work (J): The *energy* expended to oppose a *force*.

Measuring the World, by John Austin

Measuring the World, by John Austin

Appendix B: Conversion Tables

Conversions are exact when indicated by *, and are otherwise given to finite precision.

B.1 Length

Imperial → Metric		Metric → Imperial	
1 inch	25.4* mm	1 mm	0.03937008 in
1 foot	304.8* mm	1 m	1.09361330 yd
1 yard	914.4* mm	1 km	0.62137119 mile
1 rod	5.0292* m		
1 chain	20.1168* m		
1 furlong	201.168* m		
1 mile	1609.344* m		
1 nautical mile	1852* m		

B.2 Area

Imperial → Metric		Metric → Imperial	
1 sq in	6.4516* cm^2	1 cm^2	0.15500031 in^2
1 sq ft	0.09290304* m^2	1 m^2	1.19599005 yd^2
1 sq yd	0.83612736* m^2	1 m^2	10.7639104 ft^2
1 acre	0.40468564 ha	1 ha	2.47105381 acre
1 sq mile	2.58998811 km^2	1 km^2	0.38610408 mile^2

B.3 Volume
Liquid Measure (UK)

Imperial → Metric		Metric → Imperial	
1 fl oz	28.4130625* mL	1 mL	0.035195080 fl oz
1 pint	568.26125* mL	1 L	1.759753986 pint
1 gallon	4.54609* L	1 m³	219.9692483 gallon
1 barrel (oil)	158.987295* L	1 m³	1.383564946 barrel

Liquid Measure (USA)

Imperial → Metric		Metric → Imperial	
1 fl oz	29.5735295625* mL	1 mL	0.0338140227 fl oz
1 pt	473.176473* mL	1 L	2.1133764189 pint
1 qt	0.946352946* L	1 L	1.0566882094 qt
1 gal	3.785411784* L	1 m³	219.9692483 gallon

Dry Measure (USA)

Imperial → Metric		Metric → Imperial	
1 pt	550.6104713575* mL	1 L	1.8161659685 pt
1 qt	1.101220942715* L	1 L	0.9080829843 qt
1 gal	4.40488377086* L	1 L	0.2270207461 gal
1 pk	8.80976754172* L	1 m³	113.5103730 pk
1 bu	35.23907016688* L	1 m³	28.37759326 bu

Regular measurements

Imperial → Metric		Metric → Imperial	
1 cu in	16.387064* mL	1 L	61.023744095 cu in
1 cu ft	28.316846592* L	1 L	0.0353146667 cu ft
1 cord	3.624556363776* m³	1 m³	0.2758958338 cord

Note: 1 m³ = 1000 L --> 1 m³ = 35.31 cu ft.

B.4 Mass

Imperial → Metric		Metric → Imperial	
1 grain	64.79891* mg	1 mg	0.01543235835 grain
1 dram	1.7718451953125 * g	1 g	0.56438339119 dram
1 oz	28.349523125* g	1 g	0.03527396195 oz
1 lb	453.59237* g	1 kg	2.20462262185 lb
1 stone	6.35029318* kg	1 kg	0.15747304442 stone
1 cwt	50.80234544* kg	1 kg	0.01968413055 cwt
1 ton (short)	0.90718474* t	1 t	1.1023113109 short ton
1 ton (long)	1.0160469088* t	1 t	0.9842065276 long ton

B.5 Temperature

$°F = 32 + 1.8 \, °C$
$°C = (°F - 32) \times 5/9$
$K = °C + 273.15$

B.6 Other Quantities

Imperial → Metric   Metric → Imperial

	Imperial		Metric	Metric	Imperial
Force	1 lbf	4.4482216152605 * N	1 N	0.2248089431 lbf
	1 pdl	0.138254954376* N	1 N	7.23301385 pdl
Pressure	1 psi	6894.75729 Pa	1 hPa	0.01450377378 psi
	1 psi	51.71493255 Torr	1 Torr	0.01933677471 psi
	1 in Hg	3,386.38816 Pa	1 hPa	0.02952998749 in Hg
	1 Torr	133.3223684 Pa	1 hPa	0.7500616828 Torr
	1 atm	1.01325×10^5* Pa	1 atm	14.6959488 psi
Energy	1 BTU	1.0543502645 kJ	1 kJ	0.9484514147 BTU
	1 therm	29.28750735 kWh	1 kWh	0.03414425093 therm
	1 ton TNT	4.184×10^9*J	1 GJ	0.239006 ton TNT
Power	1 hp#	745.6998716 W	1 kW	1.3410220896 hp

Mechanical horse power, based on 1 hp = 550 ft lbf s^{-1} exactly.

Measuring the World, by John Austin

Appendix C: Special SI units named in honour of scientists

ampere   A   Electrical Current
André Marie Ampere, 1775-1836. French mathematician and physicist. Showed that currents in a wire produce a magnetic field and two parallel wires with currents generate a force between them.

becquerel   Bq   Radioactive decay rate
Antoine Henri Becquerel, 1852-1908. French physicist and engineer, discoverer of radioactivity along with Marie and Pierre Curie. Discovered spontaneous radioactivity in Uranium salts.

Celsius   °C   Temperature
Anders Celsius, 1701-1744. swedish astronomer. Was the first to link aurorae in the atmosphere with the Earth's magnetic field. Developed the centigrade (later Celsius) scale in 1742.

coulomb   C   Electrical Charge
Charles Augustin de Coulomb, 1736-1806. French physicist, inventor and army engineer. Made fundamental contributions to friction, electricity and magnetism. Formulated Coulomb's law concerning the forces between charges

farad   F   Electrical capacitance
Michael Faraday, 1791-1867. English chemist and physicist. Discovered the principle of electromagnetic induction in 1831: moving a magnet through a coil of copper wire induces a current in the wire.

gray   Gy   Absorbed radiation dose
Louis Harold Gray, 1905-1965. British physicist who worked on the effects of radiation on biological systems. In 1937, built an early neutron generator and used it to study the biological effects of neutrons.

Measuring the World, by John Austin

henry	H	Electrical Inductance
		Joseph Henry, 1797-1878. American physicist. Improved the electromagnet. Invented the telegraph and discovered electromagnetic induction independently of Samuel F.B. Morse and Michael Faraday.
hertz	Hz	Frequency
		Heinrich Rudolph Hertz, 1857-1894. German physicist, discovered electromagnetic waves between 1886 and 1888. Using an oscillating electric spark, he showed that similar oscillations occurred in a distant wire loop.
joule	J	Work, energy
		James Prescott Joule, 1818-1889. British physicist, helped prove the law of energy conservation. In 1840 discovered the relationship between the energy of an electrical circuit and the heat produced.
kelvin	K	Thermodynamic temperature
		William Thomson (Lord Kelvin), 1824-1907. British physicist and electrical engineer. Introduced the thermodynamic temperature scale, and was one of the first scientists to use scientific principles to determine the age of the Earth.
newton	N	Force
		Sir Isaac Newton, 1642-1727. English physicist, mathematician and astronomer, carried out pioneering work on the processes of dynamics and gravitation and a wide range of other fields. Best known for the relationship between forces and masses $F = ma$.
ohm	Ω	Electrical resistance
		Georg Simon Ohm, 1789-1854. German physicist. Described the law of electrical currents (Ohm's law). Originally discovered in 1827, the law was neglected until 1833.

Measuring the World, by John Austin

pascal   Pa   Pressure
Blaise Pascal, 1623-1662. French physicist, mathematician and philosopher: best known for his experiments with fluids and for his work on probability theory. Completed fundamental work on the pressure in fluids.

siemens   S   Electrical conductance
Verner von Siemens, 1816-1892. German electrical engineer who established the Siemens company. Builder of the first electrical elevator in 1880 and the trolley bus in 1882.

sievert   Sv   Effective radiation dose
Rolf Maximilian Sievert 1896-1966. Swedish medical physicist who studied the biological effects of radiation. He played a pioneering role in the measurement of radiation doses in the treatment of cancer.

tesla   T   Magnetic field strength
Nikola Tesla, 1856-1943. Croatian electrical engineer and inventor. Best known for the development of AC systems respnsible for mdern power generation. Also invented the Tesla coil, a high frequency transformer.

volt   V   Electromotive force
Alessandro Volta, 1745-1827. Italian Inventor and physicist, made discoveries in electrostatics, meteorology and pneumatics. Invented the voltaic pile, an early type of battery.

watt   W   Power
James Watt, 1736-1819. Scottish engineer whose improved engine design made steam power practicable by using the principle of a separate condenser for steam. Also did important work in chemistry and metallurgy.

weber	Wb	Magnetic flux
		Wilhelm Eduard Weber, 1804-1891. German physicist, who with Carl Friedrich Gauss invented the electromagnetic telegraph. Helped to develop the theory of electrodynamics.

Reference :
http://en.wikipedia.org/wiki/List_of_scientists_whose_names_are_used_as_SI_units

Appendix D: Ten Units which should be Confined to the Dustbin of History

Poorly-known units have been extensively discussed in the text. Of course the whole of the fps system has had its day, but for those that continue to use it, if at least the following units could be completely abandoned, the rest of us would be grateful.

Unit name	Chapter	Remark
rod	5.5	This is probably no longer used anywhere
US survey foot	5.5	Why have a unit that differs from the real thing by less than 2 parts in a million?
peck	6.9	Is it a volume or a mass?
bushel	6.9	This has been complicated by commercial interests who want to use mass masquerading as a volume. Why?
barrel	6.9	Does anyone outside the oil industry really know what this is?
troy ounce	7.5	The ounce is a bad enough unit, why complicate it?
slug	7.5	Its use confuses forces and masses and seems to be unknown outside US engineering applications
carat/karat	7.5	Sometimes a mass, sometimes a degree of purity. Make up your mind or better still, abandon it
ton TNT	11.7	It's not even the energy produced by a ton of TNT, so why not just use GJ?
quad	12.8	Why not just use Exajoules? They're almost the same size.

Measuring the World, by John Austin

Measuring the World, by John Austin

Index

absolute zero 44,150,285,375
acceleration 24,29,32,37,131,171, 378
acidity 281
acre (unit) 13,93,115,120,373,381
age of Earth 79-80, 386
   of universe 323-324
aircraft 243-245
   energy efficiency 243
   flight level 199
   load 140-141
   maximum engine thrust 244
amount of substance 279
Ampere (scientist) 386
ampere (unit) 254, 363,376, 377
angle of incidence 295
angle of reflection 295
angular momentum 56,187
   conservation 188
approximate conversion
   area 121
   density 144
   energy 235
   force 175
   length 93
   mass 136
   volume 127
are (unit) 119
area 115-122, 382
   land 119-120
   large items 119-120
   medium items 117-119
   small items 117
   units 116

astronomical unit 320,325,329
atmosphere
   measurements 160-166
   carbon dioxide in 81,160,164,208-209
   CO_2 in (fps) 224
   geostrophic flow 202, 209
   humidity measurement 163
   ion flux 318
   mass 207-209
   mass (fps) 224
   pressure 201-207
   scale height 204
   stratosphere 161, 162
   temperature 160-162
   troposphere 161
   weather systems 162,199,212
atomic particles
   mass and energy 312-315
   nucleus size 310-311
   sizes 310-312
   weight 283, 286-287
aurora 204,268,293,386
Avogadro's constant 35,280,288, 358-359, 361, 363-364

bar (unit) 201
barometer 201
   aneroid 206-207
   mercury 205-206
barometer correction (fps) 224
barrel (unit) 126,340,370,383
barye (unit) 201
battery from fruit, vegetables 289

beat frequency 219
Beaufort scale 105
Becquerel (scientist) 386
becquerel (unit) 69, 377
bel (unit) 215
Bernoulli's theorem 189
body mass index (BMI) 140, 374
Boltzmann's constant 35, 155, 361, 363, 364
bomb, energy 240
    atom bomb 240
    H-bomb 240-241, 268
    TNT equivalent 240, 385
braking, train 185-186
    road vehicle 185-186
breathing mixtures 213-214
breeze, land 210-211
    sea 210-211
brightness 299-304
British Thermal Unit 234
bubble pressure 193
bushel (unit) 126, 370, 383
Buys Ballot's law 106, 209

caesium 133 splitting frequency 52, 56, 360, 375
calendar 46-51
    Chinese 46
    Gregorian 47-48
    Islamic 46
    Julian 46
    World 49-51
calorie (unit) 232
calorimeter 236
candela (unit) 299, 363, 376
capacitance 256, 377, 378, 385,
capacitor 256

carat (unit) 137, 389
Carbon-14 production 71, 279, 317
catalysis 289-290
celestial sphere 323
Celsius (scientist) 385
Celsius (unit) 150-153, 376
chain (unit) 89, 120
charge 34, 254, 284, 288, 360, 362, 377
chemical elements 57, 281-284
    similarities 281
climate change 81, 128, 163-166
clock 51-55
    atomic 52-54
    pendulum 52, 54-55
    sand 51
    water 51
colour vision 94
compass 103, 254, 260
conversion tables
    area 381
    length 381
    mass 383
    temperature 383
    volume 382
    other quantities 384
conversion to metric 335-343,
    cost 341-343
conversion of idioms 336-337
Coulomb (scientist) 385
coulomb (unit) 254, 377
couple 188
cord (unit) 127
Coriolis acceleration 209
cosmic rays 316-319
    effects on health 319
    effects on electronics 318-319

cubit 86

dalton (unit) 281
dating
   archaeological 71-75,78
   igneous rocks 75-76,78
   potassium 76-77,78
   radiocarbon 71-75,78
   sedimentary rocks 77,78
   strontium 78
   Turin shroud 74
   uranium 74-76,78
   with earth magnetic field 80
De Broglie wavelength 315
decibel (unit) 215
decimal currency 336, 338
declination 98,323
deep sea animals 214
density 143-145, 378
   gases 286-287
   highest and lowest 144-145
diffraction 296-298
distance of the moon 319-320
   of the sun 320
dual units 340-341, 344-347
   problems with 344-354
   Gimli glider 344-347
   Loss of Korean Air cargo 347-348
   Mars climate Observer 347
dyne (unit) 175

Earth, circumference 90
   diameter 90
   mass 180
   precession 190
   tilt of axis 190

earthquakes 224-225
electrical
   current 254-255,375,376,378
   capacitance 256,377,378
   conductance 255,377
   conductivity 255-256
   inductance 256-257
   multimeter 258
   resistance 255,377,378
   voltage 255,379,387
electric field 254,362
electricity 253-272
   animal 264
electromagnetic spectrum 94-98, 321
electromagnetism 36,172,173,362
electron charge 35, 362
electron volt (unit) 312
energy 231-248, 378
   bombs 240
   dietary 235
   dietary: adult daily need 235
   earthquake 242
   efficiency in home 271-272
   fuels 244-245
   ethanol and petrol (gasoline) mix 245
   geothermal 270
   heat 233
   hurricane 242
   mechanical 233
   tropical storm 242
   volcanic eruption 242
entropy 44
erg (unit) 233
escape velocity 177,200

exercise energy 237-240
    walking and running 238-239
    cycling 239-240
    human flight 240
explosive boiling 193
exponential decay 69
exponential growth 17
extinction of dinosaurs 77, 242

Fahrenheit (unit) 150-153
farad (unit) 256, 377
Faraday (scientist) 385
firefly 293
fireworks 244
floating insects 193
flow along a pipeline 192
fluid ounce (unit) 123-124
foot (unit) 86-87, 89
foot-candle (unit) 300
force 171-193, 376
    electromagnetism 172-174
    strong nuclear 172-174
    weak nuclear 172-174
    between two magnets 261-262
fps units
    dependent on metric 368
    inconsistency of 368
fracking 221
Fraunhofer lines 323
frequency 67-82, 376, 378, 386
friction 156, 185-187
    static 185
    rolling 185
fuel production 265
fundamental constants 35
    SI 359-363

furlong (unit) 89, 120, 381

galaxy size 321
galvanometer 258
gamma rays 68, 95, 321
gas chromatography 284
    narcosis 213-214
    properties 153-154
gauss (unit) 260
geysers 270
Gimli glider 344-345
global positioning system (GPS) 101
grain (unit) 136, 383
gravitational constant 35, 177
gravity 172-173, 176-177, 178-180, 181-185
    measuring 178-180
    Newton's law 177, 180
gravity assist 183
Gray (scientist) 385
gray (unit) 70, 376
greenhouse effect 160, 163-166
Greisen Zatsepin Kuzmin (GZK) limit 317-318
gyroscope 190

half-life 69
hardness 192-192
hectare (unit) 119
Heisenberg uncertainty principle 297, 311
helicopter flight 189
Henry (scientist) 392
henry (unit) 256, 377, 386
Hertz (scientist) 386
hertz (unit) 68, 376

Measuring the World, by John Austin

Hertzsprung-Russell diagram 328
Higgs boson 313
horse power (unit) 237
Hubble constant 324
Hubble space telescope 319, 322
hurricane 105-107
hydrostatic approximation 217

ice skating 187, 188
ideal gas 285-288
imperial units 11, 23, 26
inch 87, 89
infra-red 95, 322
International Avogadro Project 357-358
International Prototype Kilogramme 24, 33, 131, 134, 356-359

jet stream 107
Joule (scientist) 386
joule (unit) 233, 377, 378

karat (unit) 137, 389
katal (unit) 290, 377
Kelvin (scientist) 386
kelvin (unit) 154, 375, 376
Kepler's laws 176, 326
kilogramme (unit) 131-134, 356-359, 376, 378
kinetic energy 149, 312
Kleiber's scaling law 142
knot (unit) 104

lamp kerosene 304
league 89

leap, day 47, 48
  year 47
  second 57-58
length 85-110, 378
Leonardo da Vinci Vitruvian man 88
lepton 173, 313
light 96, 295-305, 324
  effects on insects 298
light bulbs 271-272, 301-305
  compact fluorescent 302
  efficacy 303
  efficiency 303
  halogen 302
  incandescent 271-272, 301-302
  metal halide 301-302
  sodium 301
light emitting diodes (LEDs) 302-305
light levels 304-305
light meter 300-301
light year (unit) 320
lightning 263-264
litre (unit) 116, 122-124
lumen (unit) 300, 377
luminosity 324, 327, 377, 378
luminous efficacy 360
luminous flux 300, 377, 378
luminous intensity 299-300, 376
lux (unit) 300, 377

machines, simple 246
  energy efficiency 246
  inclined plane 246-247
  lever 247-248
  mechanical advantage 246
  velocity advantage 246

Measuring the World, by John Austin

Mach number 218
magnet 259
magnetic dipole 259
   field 173, 254
   field of mobile phone 262
   flux 257, 260
   force 260
   levitation 260
   moment 261
magnetic resonance imaging (MRI) 260-262
magnetism 259-263
map scaling 101-102
   projections 101
Mars climate Observer 348
mass 33-34, 131-145, 312-315, 376, 377, 383
   connection with volume 115, 124, 132
   difference from weight 134-135
mass of electron 35, 313
   neutron 35, 313
   proton 35, 313
mass primary standard 33
mass spectrometer 77, 284
Maxwell's equations 254, 363
measurement precision 368
metre (unit) 30-31, 90-91, 362, 376, 377
metric system terms 376
   unit multipliers 24, 376
metrication by nation 349-353
   Former British Common-wealth 350-351, 352
   UK 350, 352
   USA 351-352
mho (unit) 255

Michelson-Morley experiment 91
microscope 315-316
   atomic force 316
   electron 315
   optical 315
   scanning probe 316
microwave 56, 68, 95
   oven 68, 97-98
Milankovitch cycles 80, 190
mile (unit) 87, 93
mole (unit) 279, 363-364, 376, 379
   number per mole 35, 288-289, 358-359
molar volumes of gases 285-287
moment of inertia 187

napier (unit) 215
nautical mile 100, 339, 381
navigation 98-101
   latitude 98
   longitude 99-100
neutrino 313
Newton (scientist) 392
newton (unit) 135, 175, 377, 378
noctilucent clouds 204
nuclear fission 266-267
nuclear fusion 267-269, 293
   tokomak reactors 268
   International Thermonuclear Reactor (ITER) 268-269
   laser fusion 269
nuclides, table of 283
   islands of stability 283

Ohm (scientist) 386
ohm (unit) 255, 377

Measuring the World, by John Austin

paper size 117-119
   sheets per acre 122
parallax 320
parsec (unit) 321
Pascal (scientist) 387
pascal (unit) 201, 377, 378
peck (unit) 126, 389
pedometer 109
pendulum 178-180
   Foucault 179-180
periodic table of the elements 281-283
pH (unit) 289
phon (unit) 217
Planck scale 38
   units 38
Planck's constant 35, 359, 361-362, 364
plasma 268
point size (printing) 94
polar ice caps 128
popcorn price 125
pound (unit) 133
poundal (unit) 135, 174
power 236-237, 264-272, 378, 384
   hydro 276
   off-peak 271
   renewable 269-270
   solar 270
   wind 270
   worldwide production 266
power station 264-270
   clean coal 265
   efficiency 264
   nuclear 266-269
   output 266

Pressure 199-225, 377, 384
car tyre 222-223
car tyre (fps) 225
car exhaust 223
conversion to mean sea level 202-203
isobar 202
sigma levels 203
subterranean 214
surface pressure 203
variation with height 206-207
underwater 212-214
underwater (fps) 224-225
units 200
primary standard 30, 33
pyrometer 159

quantum 17, 302
quantum mechanics 37, 159, 294, 311, 312, 315
quarks 173, 313
quart (unit) 124

radian 188, 300, 377
radioactivity 68-70
   human exposure 70-71
radio broadcasting 96-97
radio wave 95, 322
rainbow 94
redshift 324
reflection 295
refraction 295-296
refractive index 296
relativity general 44, 85, 101, 174, 177
   special 44, 91-92, 254

resolving power 315
Richter scale 221-222
right ascension 323
rockets 243-245
    thrust 244
rotational force 187-191
    water in sinks and tubs 209-212

Saffir-Simpson hurricane scale
    106,107
satellites 181-184
    geosynchronous 181
    polar orbiting 181
scale analysis 37-39
scaling, biological 141-142
Schrödinger's cat 312
scientists named in metric units
    385-388
second (unit) 55-57, 375, 376,378
seismograph 221
Siemens (scientist) 387
siemens (unit) 255, 377
Sievert (scientist) 387
sievert (unit) 70, 376
sight, long 102-103
    short 102-103
SI new 355-364
    spelling of names 14-15
SI units 14,23-37,355-364, 375-377
    base 23-31,376
    derived 32-33,376-377
slug (unit) 137,340,389
Snell's law 296
solar wind 317

sound waves 215-218, 219-222
    loudness 200,215
    perception human 215-216
    perception other animals 216
    in weather forecast 217
sound waves in solids 219-222
    compression waves,
    P-waves 219-220
    shear waves, S-waves 219-220
space debris 182-183
    edge of 204
specific gravity 143, 378
specific heat capacity 233
speed 103-104, 378
speed of light 34,35,91,360
    in medium 296
speed of sound 217-218,219-222
    air (fps) 225
    gas 217-218
    liquid 218
    solid 218,219-222
    water (fps) 225
spin rate 191
sport distance measurement
    108-110
    jumps measurement 109
    throws measurement 109
    timing 59-62
    track measurement 108-109
Stars 324-329
    absolute magnitude 326
    apparent magnitude 326
    colour 328
    luminosity 327
    standard candle 324
Stefan-Boltzmann constant 35, 160
steradian (unit) 299-300, 377

Measuring the World, by John Austin

stone (unit) 135, 338
storage heaters 271
sun 324-329
    spots 324-325
sundial 51
surface tension 192-193, 378

tachometer 191
telescopes 319,321-323
    multiple 322-323
    space telescope 319,322
television 96-97
temperature 149-166
    atmospheric 160-166
    conversion 154-155
    fixed points 152-153,157
    Goldilocks effect 150
    greenhouse effect 163-166
    highest and lowest 151
    measurement 158-159,162-163
    thermodynamic 155-158
Tesla (scientist) 387
tesla (unit) 260, 377
therm (unit) 234
thermocouple 160
thermometer 150-153, 158-159
    alcohol 158
    bimetallic 160
    maximum-minimum 162
    mercury 158
    platinum resistance 159
    radiation 159-160
tides 184-185
    neap tide 184
    spring tide 184
    sun and moon effect 184-185

time 43-62, 376
    travel 44
ton, long 135,383
    short 135,383
tonne (unit) 136,383
tornado 106-107
torr (unit) 201
total solar irradiance 325
traffic flow 187
transformer 257
tsunami 221
turbulence 107-108

ultrasound 215
ultra-violet 95,98,162,325
unit agreement 335-352
unit, alternate system 37-39
    base 23,375-376
    computing 39-40
    derived 23,32-33, 376-377
    systems 26-30
    cgs system 24,26
    fps system 24,26
    mks system 24,26
    SI 25,27
    SI new 355-364
    spelling of names 14-15
units poorly known 389
universe size 321

vacuum 205
Van Allen belts 325
velocity 103-104, 378
vernier scale 205
viscosity 52,191, 379
volt (unit) 255, 377

Measuring the World, by John Austin

Volta (scientist) 387
Voltage 377
volume 116, 122-127, 379, 382
    connection with mass 123, 131
    units 116
Voyager spacecraft 183-184

Watt (scientist) 387
watt (unit) 236, 377
watt balance 358, 361
wave-particle duality 294, 297
weather forecasting 202
Weber (scientist) 388
weber (unit) 260, 377

weighing machine 138-139
    analytical balance 138, 139
    bathroom scales 138
    steelyard 138
weight 131, 134-135, 378
    aircraft baggage 140-141
    difference from mass 134-135
    troy 137, 389
Wheatstone bridge 258-259
Wien displacement law 160, 328
wind 105-108
work 231, 233, 379
worldsday 49, 50

X-rays 95, 322

yard (unit) 87, 89, 90, 381
    shrinking of 90
Young's double slit experiment 297

Measuring the World, by John Austin

John pictured at Bryce National Park, Utah, USA

About the Author

The Author, Dr. John Austin, has over 30 years' research experience on the upper atmosphere and has published over 80 papers in numerous international scientific journals. In addition John worked for 4 years as an Editor of the Journal of Geophysical Research, the premier Geophysics journal in the USA.

He has spent several years working in the USA, at NASA Langley, Hampton, virginia (1984-1985) and the University of Washington (1988-1990), where amongst other things he met his future wife Alda, to whom he is still married. During 2003-2011 John worked in Princeton, NJ, USA. His main scientific contribution has been to show the connection between ozone depletion and climate change. John has been involved in the writing of numerous international reports for the World Meteorological Organisation and The Intergovernmental Panel on Climate Change, for which the IPCC received the 2007 Nobel peace prize.

In recent years, John has broadened his work into popular science, through the website http://www.DecodedScience.com and in 2014 he created an internet scientific publishing business Enigma Scientific Publishing, http://www.enigmascientific.com. He has always had an interest in our unit system and the science of measurement, and the book "Measuring the World" is his first popular science book.

****

When not working, John enjoys a variety of activities including chess, running, photography and travel. He has become addicted to sudoku, and that will be the subject of his next book!

www.ingramcontent.com/pod-product-compliance
Lightning Source LLC
Chambersburg PA
CBHW071354170526
45165CB00001B/35